Groundworks®

Algebraic Thinking

4

Carole Greenes • Carol Findell

Acknowledgments

Dr. Carole Greenes
is a professor of Mathematics Education at Boston University. She has written and collaborated on more than 200 publications, and her work has brought teachers and students closer to the NCTM Standards. Dr. Greenes was recently inducted into the Massachusetts Mathematics Educators Hall of Fame.

Dr. Carol Findell
is a Clinical Associate Professor of Education at Boston University. She has served as a mathematics educator for more than 30 years, during which time she has led writing teams for national mathematics competitions. Dr. Findell is a well-respected author and is a frequent speaker at national mathematics conferences.

Cover and interior illustrations by John Haslam.

www.WrightGroup.com

Printed in the United States of America.

Send all inquiries to:
Wright Group/McGraw-Hill
P.O. Box 812960
Chicago, Illinois 60681

ISBN: 1-4045-3193-9

1 2 3 4 5 6 7 8 9 MAL 11 10 09 08 07 06 05

The McGraw-Hill Companies

Contents

Why Teach *Algebraic Thinking* to Your Students?

Algebra, for all students at all grade levels, is being promoted by the National Council of Teachers of Mathematics (NCTM) in their *Principles and Standards for School Mathematics* (2000); by the College Board's Equity Project; and by authors of the *SCANS Report* (1991), which identifies needs of workers in the twenty-first century. As a result, school districts throughout the country are requiring all students to enroll in a formal course in algebra, usually Algebra I, no later than grade 8. In some districts, students are taking Algebra I in grades 6 and 7. Although students are capable of succeeding in algebra, they often do not. Why aren't they successful?

From the authors' experience, even more able students, jumping from an arithmetic-based program directly into the study of algebra, often find the new content confusing. The main reason for this difficulty most likely stems from lack of preparation. Although the NCTM has recommended that students gain experience with the big ideas of algebra during their elementary school years, current mathematics programs do not include such a preparatory program.

Bibliography

Greenes, Carole, Carol Findell, Mary Cavanagh, Linda Dacey, and Marion Small. *Navigating through Algebra in PreKindergarten – Grade 2. Principles and Strands for School Mathematics Navigation Series.* Reston, VA: National Council of Teachers of Mathematics, 2001.

Moses, Barbara. *Algebraic Thinking, Grades K–12: Readings from NCTM's School-based Journals and other publications*. Reston, VA: National Council of Teachers of Mathematics, 1999.

National Council of Teachers of Mathematics. *Principles and Standards for School Mathematics.* Reston, VA: The Council, 2000.

Secretary's Commission on Achieving Necessary Skills. *What Work Requires of Schools.* SCANS Report of America 2000. Washington D.C.: U.S. Department of Labor, 1991.

What Are the Six Big Ideas of *Algebraic Thinking*?

Groundworks: Algebraic Thinking for grades 1–7 provides challenging development of six big ideas in algebra. The problems build on students' experiences with arithmetic reasoning and help them make the connection between arithmetic and algebra. The six big ideas are:

- Representation
- Balance
- Function
- Proportional Reasoning
- Variable
- Inductive Reasoning

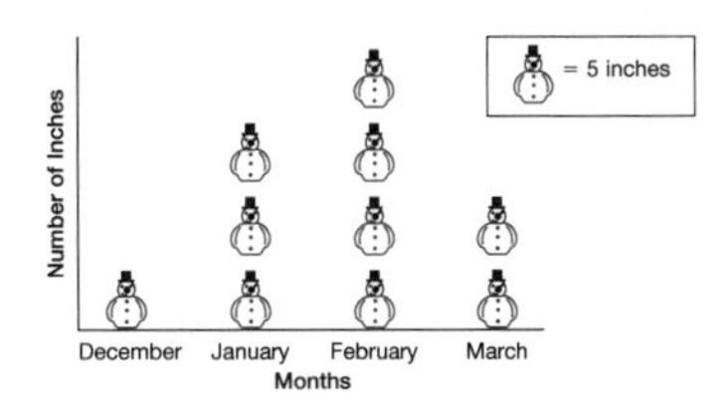

Representation

Representation shows mathematical relationships in various displays. Students read, compare, and interpret information, and explain their reasoning. Students also match or create different representations of the same relationship.

Proportional Reasoning

Students reason proportionally when they make trades, compute unit costs, interpret maps and scale drawings, make enlargements and reductions, generate equivalent ratios, and identify percentages or parts of groups.

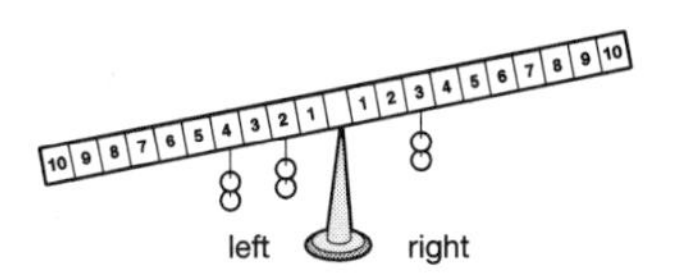

Balance

Students develop understanding of equality and of ways to modify inequalities to achieve balance or equality. Students work with balances to model one, two, or three equations with one, two, or three unknowns.

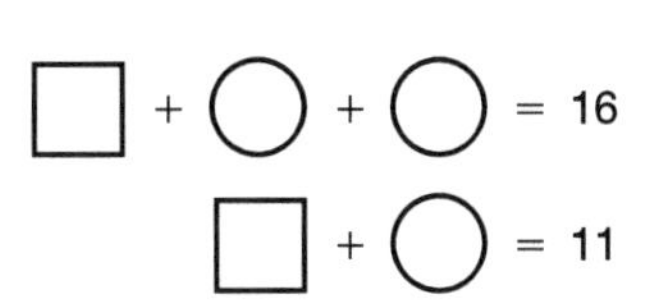

Variable

Variables are used to represent unknowns in equations, or to represent quantities that vary, as in formulas and functions. Students learn that same shapes represent the same value, and that different shapes represent different values. They identify relationships between variables and numbers and solve for the values of one, two, or three unknowns.

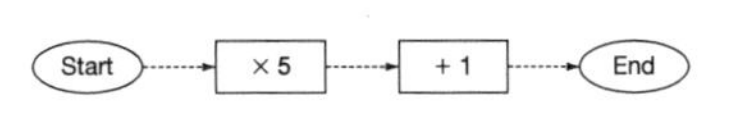

Function

Students learn to describe rules that relate inputs and outputs in function machines. They use the rules to predict outputs and to determine inputs. Students also learn to use symbols to represent the rules for relating inputs to outputs and outputs to inputs.

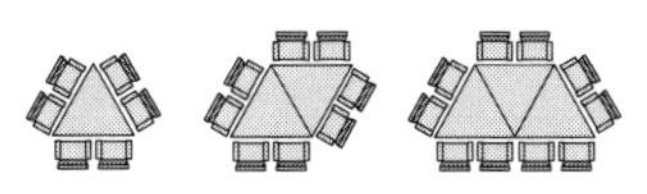

Inductive Reasoning

Reasoning inductively involves the ability to examine particular cases, identify patterns and relationships among those cases, and extend the patterns and relationships. Students also learn to represent the generalizations using words or symbols.

What is in This Book?

This book contains:

- 15 blackline-master problem sets (90 pages of problems)
- solutions for all problems
- specific teaching suggestions and ideas for each problem set
- general teacher information

Problem Sets

Each problem set consists of eight pages. The first page presents teaching information, including goals listing specific mathematical reasoning processes or skills, questions to ask students, and solutions to the first problem. The next six pages are all reproducible problem pages. The first problem is a teaching problem and is of moderate difficulty. The remaining five problems in the set range from easier to harder. Solutions to all problems are given on the eighth page of each problem set. For most problems, one solution method is shown; however, students may offer other valid methods. The mathematics required for the problems is in line with the generally approved mathematics curriculum for the grade level.

How to Use This Book

Because many of the problem types will be new to students, you may want to have the entire class or groups of students work on the first problem in a set at the same time. You can use the questions that accompany the problem as the basis for a class discussion. As the students work on the problem, help them with difficulties they may encounter. Students are frequently asked to explain their thinking. You may choose to do this orally with the whole class. After students have several experiences telling about their thinking and hearing the thinking of others, they are usually better able to write about their own thinking. Once students have completed the first problem in a set, you may want to assign the remaining problems for students to do on their own or in pairs in class or for homework. If students have difficulty with the first problem in the set, you might do more of the problems with the whole class.

Although the big ideas and the families of problems within them come in a certain order, students need not complete them in this order. Students might work the problem sets based on the mathematical content of the problems and their alignment with your curriculum, or according to student interests or needs.

There is a Management Chart that you may duplicate for each student to keep in a portfolio. You may award the Certificate of Excellence upon the successful completion of the problem sets for each big idea.

Management Chart

Name ______________________ Class ____________ Teacher ______________________

BIG IDEA	PROBLEM SET				DATE
■□□□□□ Representation	In the Graph	1	2	3	
		4	5	6	
	Who Is It?	1	2	3	
		4	5	6	
■■□□□□ Proportional Reasoning	Bead Strings	1	2	3	
		4	5	6	
	Better Buy	1	2	3	
		4	5	6	
	Creature Trades	1	2	3	
		4	5	6	
■■■□□□ Balance	Balance Beam	1	2	3	
		4	5	6	
	Pan Balances	1	2	3	
		4	5	6	
■■■■□□ Variable	Weighing Blocks	1	2	3	
		4	5	6	
	Missing Numbers	1	2	3	
		4	5	6	
	Grid Sums	1	2	3	
		4	5	6	
	Bridge Sums	1	2	3	
		4	5	6	
■■■■■□ Function	Function Tables	1	2	3	
		4	5	6	
	Start to End	1	2	3	
		4	5	6	
■■■■■■ Inductive Reasoning	Pattern Puzzles	1	2	3	
		4	5	6	
	Lattice Logic	1	2	3	
		4	5	6	

In the Graph

Goals

- Interpret a pictograph or bar graph.
- Match mathematical relationships presented in words with those shown in a graph.

Notes

Since the graphs in this problem set involve multiples of 2, 5, and 10, you may want to review multiplication facts with 2, 5, and 10 prior to assigning the problems.

Solutions to all problems in this set appear on page 7.

In the Graph 1

Questions to Ask

- What does each snowman represent? (5 inches of snow)
- How many inches of snow fell in January? (15 inches)
- Which month had the greatest amount of snowfall? (February)
- How can you tell from the pictograph? (There are more snowmen above February than above any other month.)

Solutions

1. February
2. December
3. It has the least number of snowmen.
4. January and February
5. March

Name ____________________

Representation

In the Graph 1

The graph shows the amount of snowfall in a city.

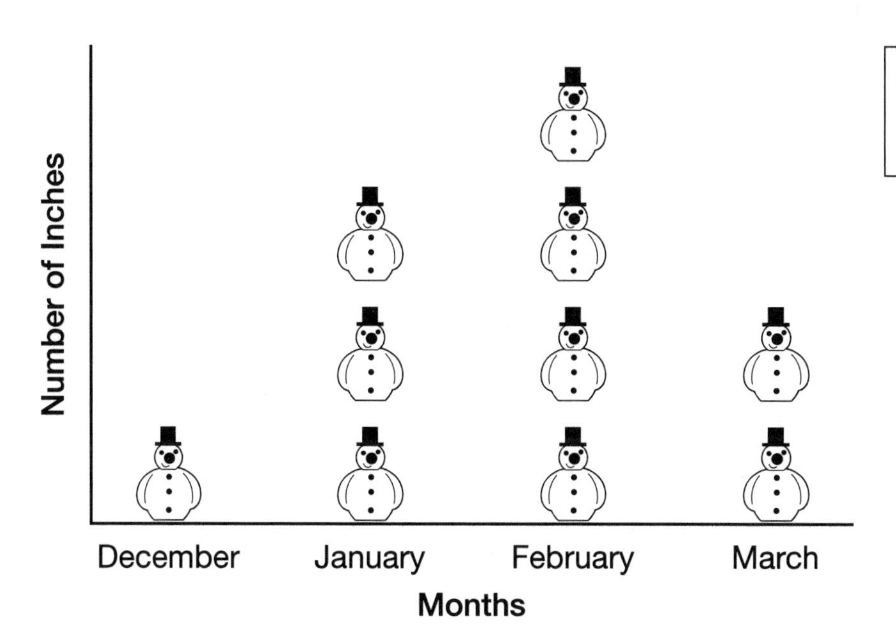

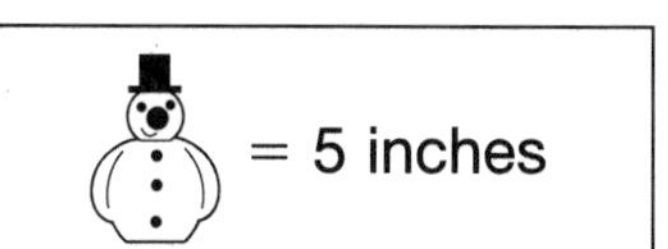

1. In which month did 20 inches of snow fall?

____

2. In which month did the least amount of snow fall?

____

3. How can you tell? ____________________

____

4. When did more than 10 inches of snow fall?

____

5. Which month had half as much snow as February?

____

In the Graph 2

The graph shows the amount of snowfall in a small town.

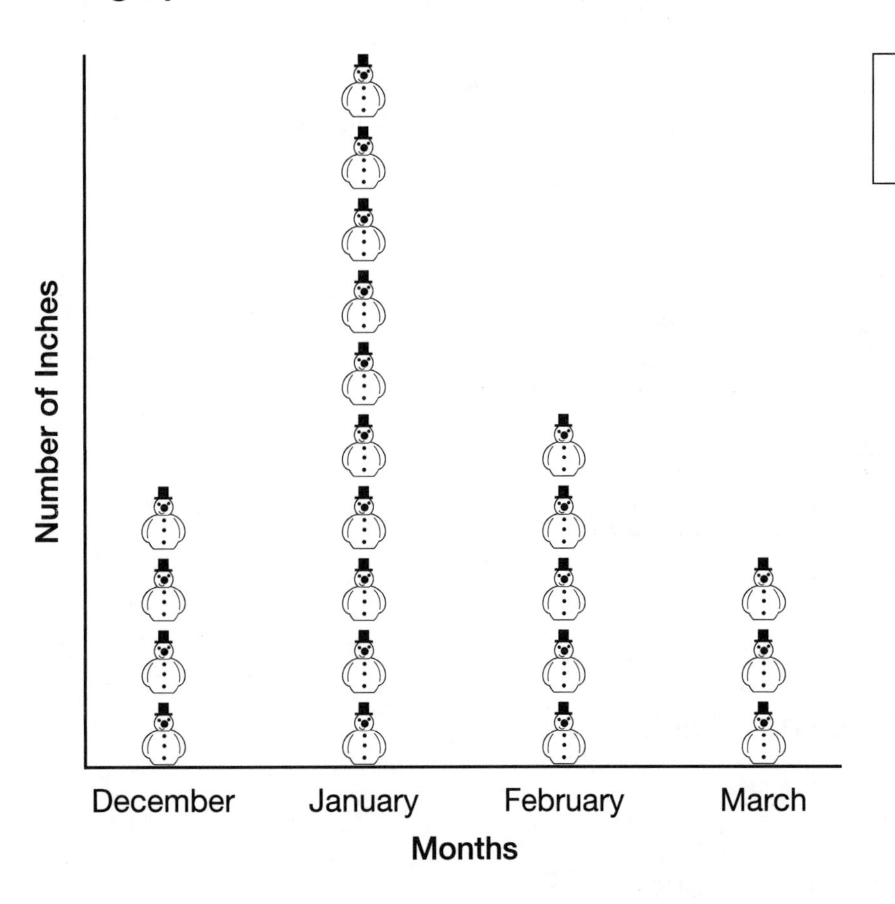

1. Which month had half as much snow as January? ____________

2. Which month had 2 fewer inches of snow than December? ____________

3. Which two months had a total of 18 inches of snow?

____

4. How many more inches of snow fell in February than in March? ____________

5. Describe two ways to find the answer. ____________________

____

Name ______________________________

■□□□□□
Representation

In the Graph 3

The graph shows the amount of snowfall in the mountains.

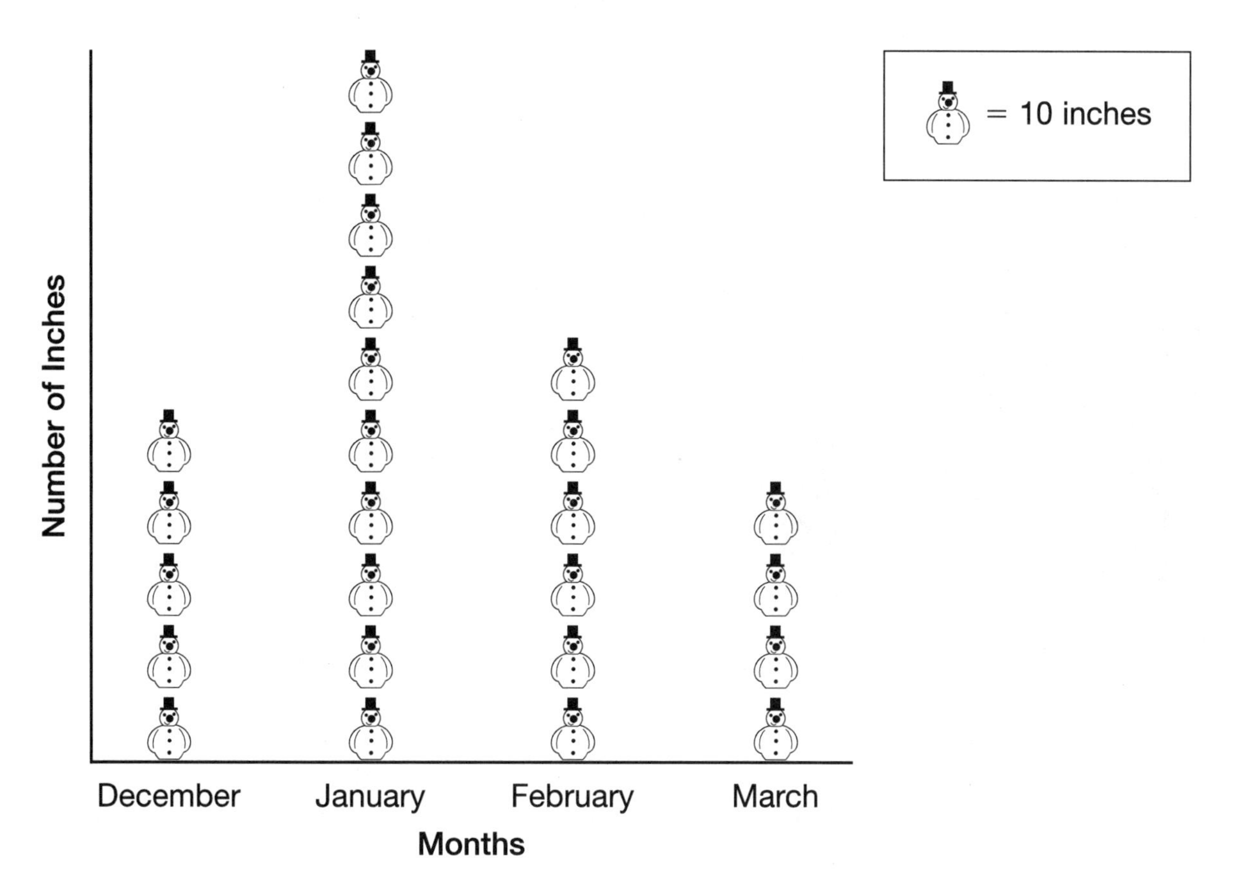

1. How much less snow fell in December than in January? __________

2. Which two months together had as much snow as January?

____

3. How many more inches of snow fell in January and February than in December and March ? __________

4. What was the total snowfall for the four months? __________

Name ____

In the Graph 4

The graph shows the number of students in three grades in a school.

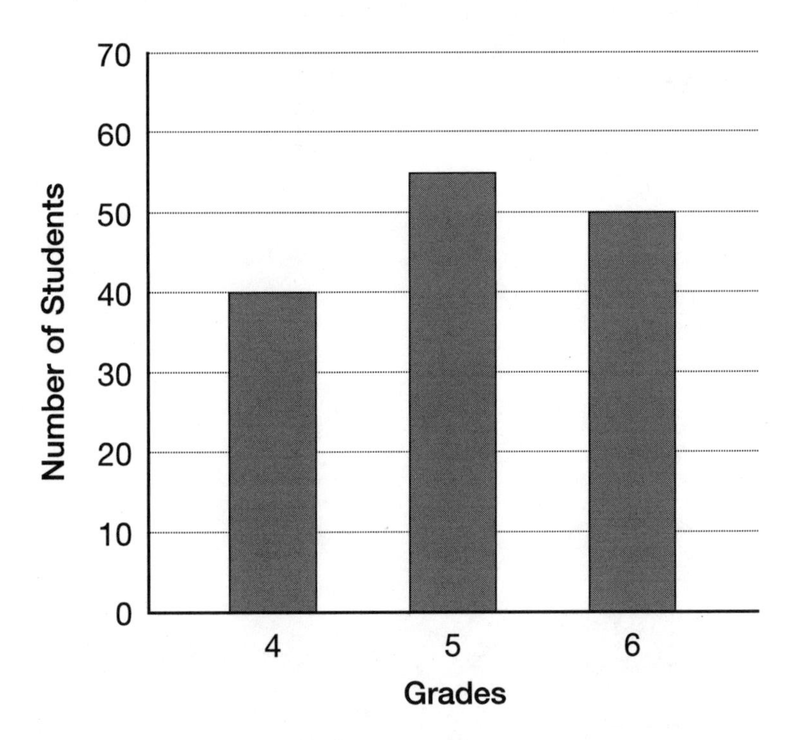

1. In which grade are there 55 students? ____________

2. How can you tell? ______________________________

____

3. How many more students are in grade 6 than in grade 4? ____________

4. Which two grades have a total of 90 students? ____________

5. What is the total number of students in grades 4, 5, and 6? ____________

Name ______________________

Representation

In the Graph 5

The graph shows the number of students in three grades in a school.

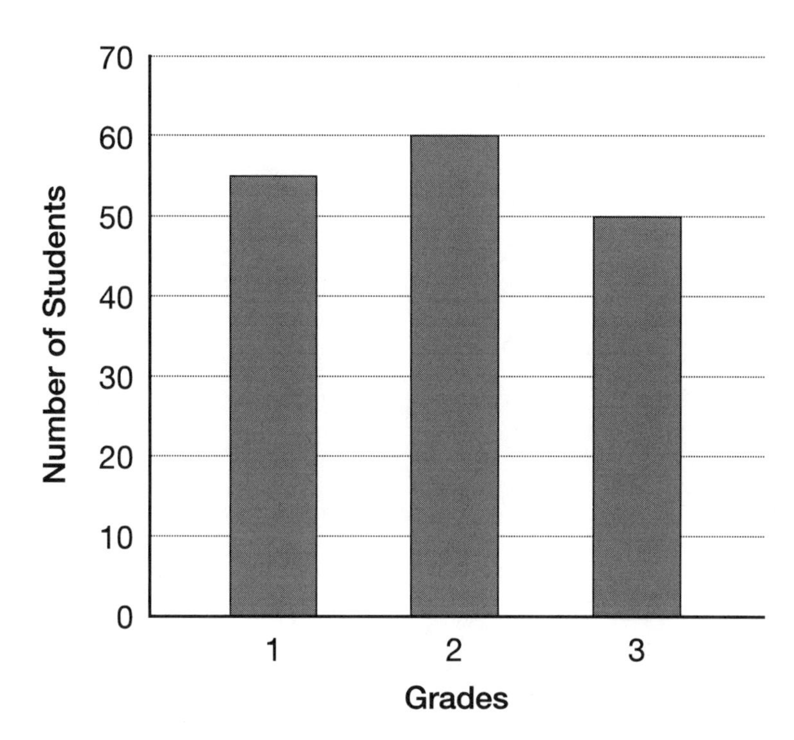

1. How many second-graders attend the school? __________

2. Which grade has five more students than grade 3? __________

3. How can you tell? ______________________________

____

4. There are eight more students in kindergarten than in grade 3.

   How many kindergarten students are there? __________

5. What is the total number of students in grades 1, 2,

   and 3? __________

Name ______________________

■□□□□□
Representation

In the Graph 6

The graph shows the number of students in four grades in a school.

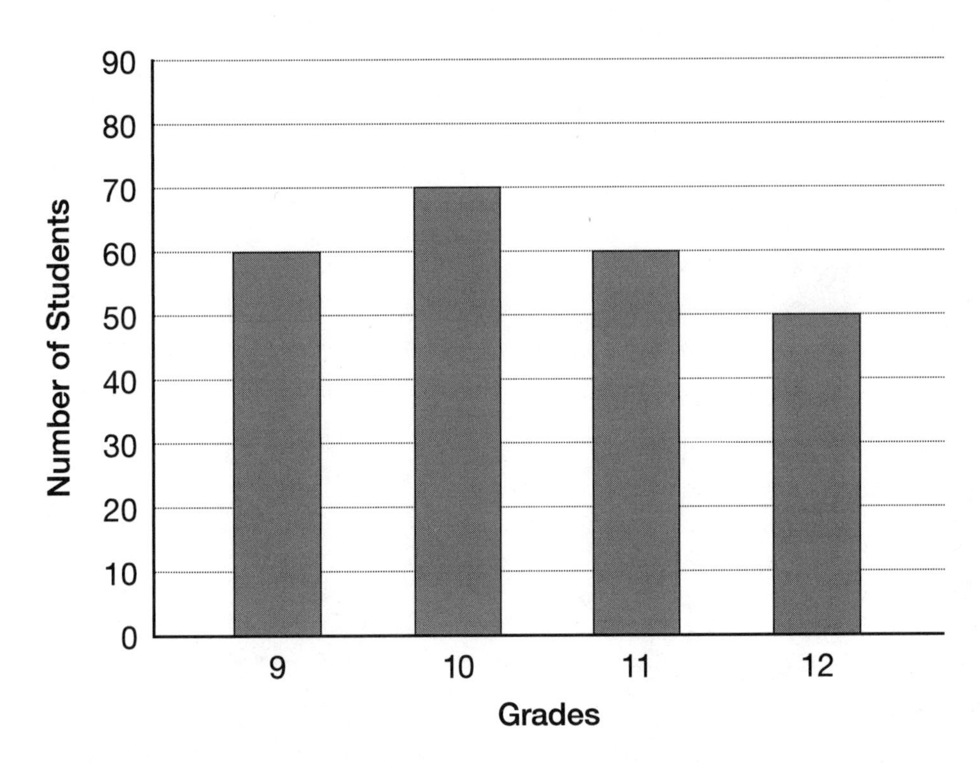

1. Which two grades have the same number of students? ____________

2. There are four classes in the 11th grade. Each has the same number of students. How many students are in each class? ____________

3. How many more students are in grades 9 and 10 than in grades 11 and 12? ____________

4. Tell two ways to find the answer. ______________________

____

5. If the total number of students stayed the same, but each grade had the same number of students, how many would be in each grade? ____________

Solutions

In the Graph 1

1. February
2. December
3. It has the least number of snowmen.
4. January and February
5. March

In the Graph 2

1. February
2. March
3. December and February
4. 4 inches
5. Possible answers: Count the number of snowmen for each month, then multiply these numbers by 2 and find the difference in the products; find the difference in the number of snowmen for the two months, then multiply this difference by 2.

In the Graph 3

1. 50 inches
2. February and March
3. 70 inches
4. 250 inches

In the Graph 4

1. grade 5
2. The grade 5 bar ends halfway between grid lines 50 and 60.
3. 10
4. grades 4 and 6
5. 145

In the Graph 5

1. 60
2. grade 1
3. The bar for grade 1 is half a space taller than the bar for grade 3.
4. 58
5. 165

In the Graph 6

1. grades 9 and 11
2. 15
3. 20
4. Possible answers: Grades 9 and 11 have the same number of students, so compare only grades 10 and 12. Grade 10 has 70 − 50, or 20, more students than grade 12; grades 9 and 10 have 130 students. Grades 11 and 12 have 110 students. There are 130 − 110, or 20, more students in grades 9 and 10.
5. 60

Who Is It?

Goals

- Interpret a scatter plot.
- Match mathematical relationships given in words with those shown in a graph.
- Make inferences.

Notes

Prior to doing this problem set, review the naming of points with *x* and *y* values on a coordinate grid. In a scatter plot, each point has two values, one from the horizontal axis (the *x*-value) and the other from the vertical axis the (the *y*-value). This will facilitate understanding of the scatter plots in the problems.

Solutions to all problems in this set appear on page 15.

Who Is It? 1

Questions to Ask

- What do the points on the graph represent? (Each point represents the number of brothers and the number of sisters for a child.)
- What does the horizontal axis show? (number of sisters)
- How many sisters does child *A* have? (0)
- Which two points represent children who have the same number of sisters? (*B* and *C*)
- Which two points represent children who have no brothers? (*A* and *D*)

Solutions

1. Barry
2. Ted
3. Janet
4. Iris

Who Is It? 1

The graph shows the number of brothers and sisters for four children. Use the clues to identify each child.

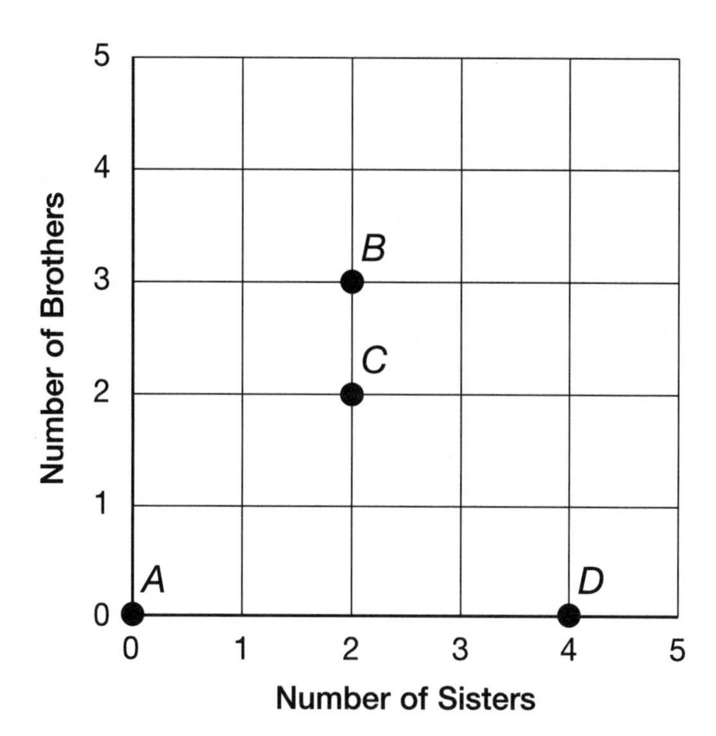

Clues

1. Barry is the only child in his family.
2. Ted and Janet have the same number of sisters.
3. Iris does not have any brothers.
4. Ted has more brothers than Janet.

1. Who does *A* represent? ______________________
2. Who does *B* represent? ______________________
3. Who does *C* represent? ______________________
4. Who does *D* represent? ______________________

Name ___

Who Is It? 2

The graph shows the number of brothers and sisters for four children. Use the clues to identify each child.

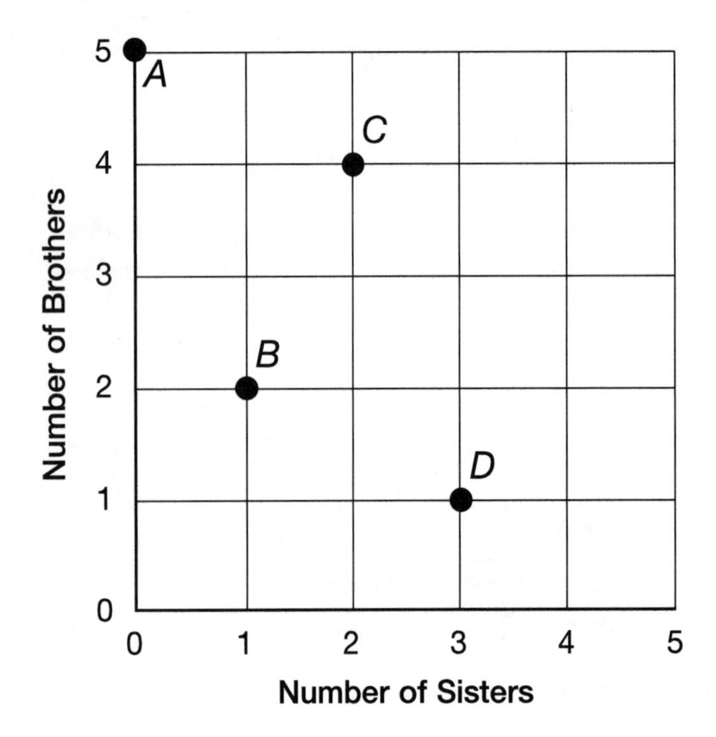

Clues

1. Eric has the fewest sisters.
2. Matt has the fewest brothers.
3. Diane has one more sister than Josh.

1. Who does *A* represent? ___

2. Who does *B* represent? ___

3. Who does *C* represent? ___

4. Who does *D* represent? ___

Who Is It? 3

The graph shows the number of brothers and sisters for four children. Use the clues to identify each child.

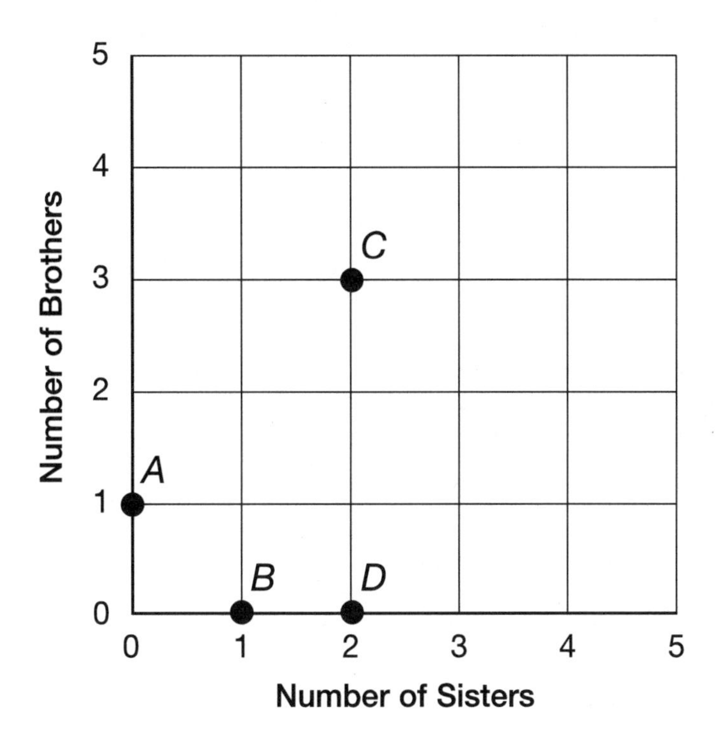

Clues

1. Angela and Larry together have four brothers.
2. Sid and Beth together have no brothers.
3. Beth has one more sister than Larry.
4. Sid has two more sisters than Larry.

1. Who does *A* represent? ___

2. Who does *B* represent? ___

3. Who does *C* represent? ___

4. Who does *D* represent? ___

Name ______________________

Who Is It? 4

Jolene, Kendra, Lee, and Don all walk home from school.
Points *A, B, C,* and *D* represent their houses.
Use the clues to find each student's house.

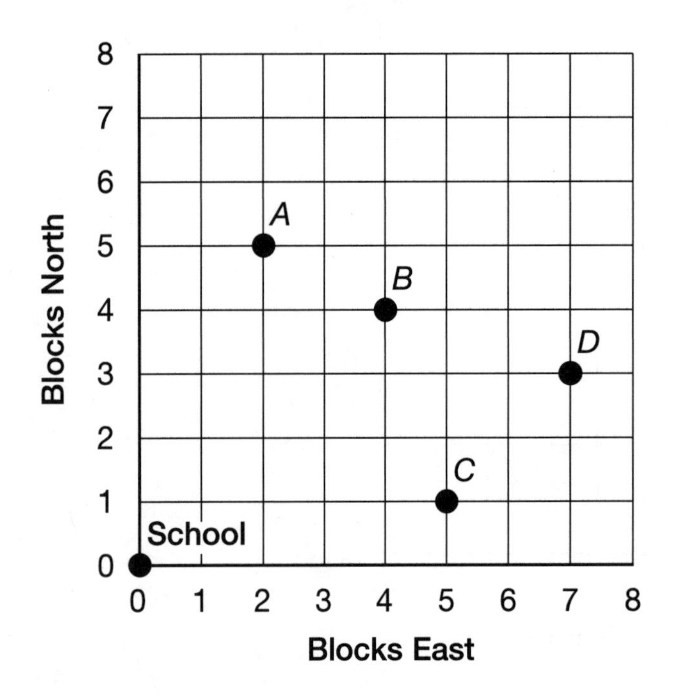

Clues

1. Jolene walks two blocks east, then straight north.
2. Kendra walks the same number of blocks east as she walks north.
3. Lee lives ten blocks from school.
4. Don lives closest to school.

1. Who lives at house *A*? ______________________
2. Who lives at house *B*? ______________________
3. Who lives at house *C*? ______________________
4. Who lives at house *D*? ______________________

Name ____________________

Representation

Who Is It? 5

Abby, Brad, Conrad, and Darlene all bike home from school. Points *A, B, C,* and *D* represent their houses. Use the clues to find each student's house.

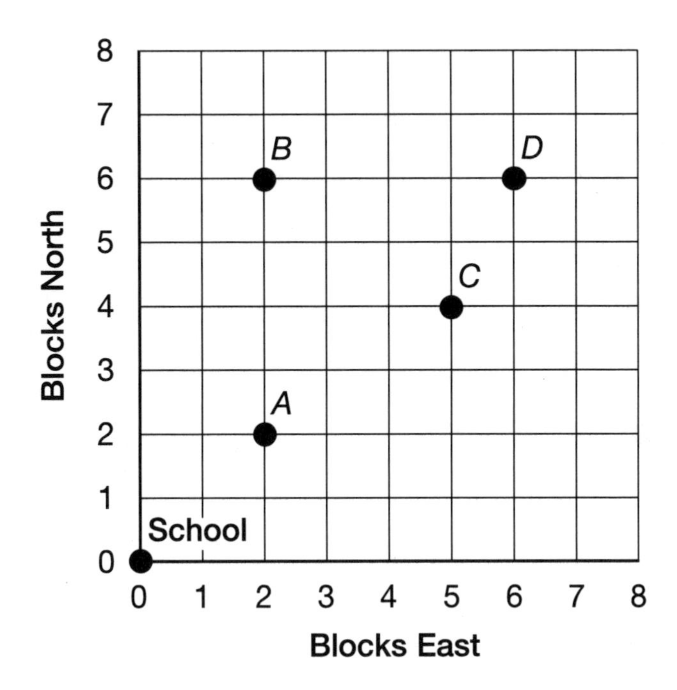

Clues

1. Brad and Darlene each live two blocks east of school.
2. Conrad bikes five blocks farther than Darlene.
3. Abby lives four blocks east of Brad.

1. Who lives at house *A*? ____________________

2. Who lives at house *B*? ____________________

3. Who lives at house *C*? ____________________

4. Who lives at house *D*? ____________________

Name ____________________

■□□□□□
Representation

Who Is It? 6

Ralph, Sonia, Tina, and Ubi all walk home from school.
Points *A, B, C,* and *D* represent their houses.
Use the clues to find each student's house.

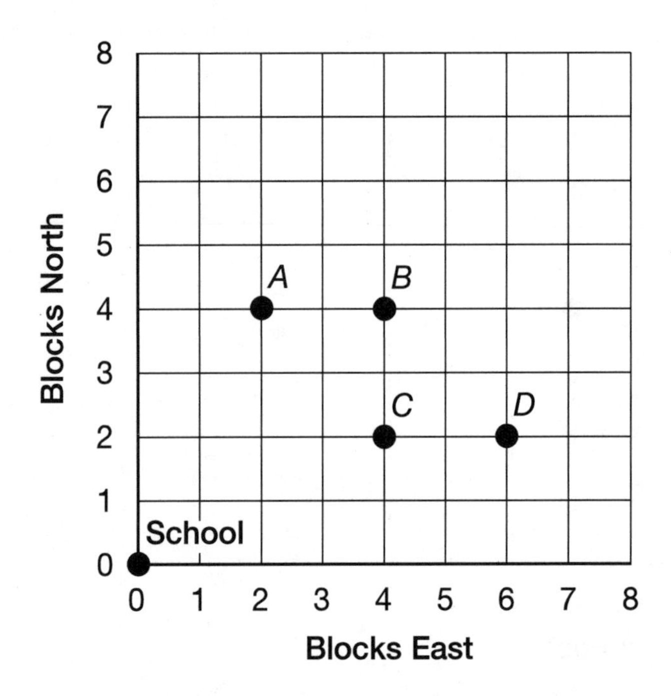

Clues

1. Sonia and Ubi each walk six blocks.
2. Ralph and Sonia each walk two blocks north.
3. Ubi and Tina each walk four blocks north.

1. Who lives at house *A*? ____________________
2. Who lives at house *B*? ____________________
3. Who lives at house *C*? ____________________
4. Who lives at house *D*? ____________________

Solutions

Who Is It? 1

1. Barry
2. Ted
3. Janet
4. Iris

Who Is It? 2

1. Eric
2. Josh
3. Diane
4. Matt

Who Is It? 3

1. Larry
2. Beth
3. Angela
4. Sid

Who Is It? 4

1. Jolene
2. Kendra
3. Don
4. Lee

Who Is It? 5

1. Darlene
2. Brad
3. Conrad
4. Abby

Who Is It? 6

1. Ubi
2. Tina
3. Sonia
4. Ralph

Bead Strings

Goals

- Generate ratios and equivalent ratios.
- Construct ratio tables to solve problems.
- Extend and generalize patterns.

Notes

To help students identify the proportional relationship between groups of different types of beads, provide them with collections of two or three different colored chips to model the bead strings in the problems. They can use the chips to extend the patterns and solve the problems. Students may also choose to make a table showing the numbers of beads for different numbers of strings.

Solutions to all problems in this set appear on page 23.

Bead Strings 1

Questions to Ask

- How many white beads are on the string? (2)
- How many black beads are on the string? (6)
- How many black and white beads are there in all? (8)
- How many white beads are needed to make two strings? (4)
- How many black beads are needed to make two strings? (12)
- How many black and white beads in all are on two strings? (16)

Solutions

1. 20 more black beads
2. 6 strings
3. 7 strings

Bead Strings 1

Jamal is making strings of beads like this one. He uses 2 white beads for every 6 black beads on each string.

1. Jamal made 5 bead strings. How many more black beads than white beads did he use? __________

2. How many bead strings could Jamal make with 12 white beads and 36 black beads? __________

3. If Jamal used 56 white and black beads in all, how many bead strings did he make? __________

Bead Strings 2

Sara made bead strings like this one. She used 2 black beads for every 8 white beads on each string.

1. How many beads would she need to make

   6 strings? __________

2. Tell how you know. ______________________________

   ____

3. Sara used 10 black beads. How many bead strings

   did she make? __________

4. How many white beads did she use? __________

Bead Strings 3

Nancy used black and white beads to make bead strings. She used 4 white beads for every 6 black beads on each string.

1. If Nancy used 24 black beads, how many white beads did she use? __________

2. If Nancy used a total of 50 beads, how many strings did she make? __________

3. How many strings could Nancy make with 20 white beads and 24 black beads? __________

4. Tell how you know. ____________________________________

____

____

Bead Strings 4

Khalil has 20 white beads and 40 black beads.

1. How many bead strings like this can he make? ________

2. How many beads of each color will be left? ________

Continuing the pattern, Khalil makes a string 42 beads long.

3. How many beads are white? ________

4. How many beads are black? ________

5. Tell how you figured out the numbers. ________________

____

____

Bead Strings 5

Sasha has 27 black beads and 20 white beads.

1. How many bead strings like this can she make? ___
2. How many beads of each color will be left? ___

Continuing the pattern, Sasha makes a string 28 beads long.

3. How many beads are white? ___
4. How many beads are black? ___
5. Tell how you figured out the numbers. ___

___

___

___

Bead Strings 6

Felipe has 6 striped beads, 10 black beads, and 14 white beads.

1. How many bead strings like this can he make? __________

2. Tell how you know. ______________________________

____

Continuing the pattern, Felipe makes a string 60 beads long.

3. How many beads are striped? __________

4. How many beads are white? __________

5. How many beads are black? __________

6. Tell how you figured out the numbers. __________

____

____

____

Solutions

Bead Strings 1

1. 20 more black beads
2. 6 strings
3. 7 strings

Bead Strings 2

1. 60 beads
2. Possible answer: There are 2 black and 8 white, or 10 beads, on each string. Sara would use 6×10, or 60, beads to make 10 bead strings.
3. 5 strings
4. 40 white beads

Bead Strings 3

1. 16 white beads
2. 5 strings
3. 4 strings
4. Possible answer: There are 4 white and 6 black beads on each string. With 20 white beads, Nancy can make $20 \div 4$, or 5, strings. She can make $24 \div 6$, or 4, strings with 24 black beads. Nancy has enough black beads for only 4 strings.

Bead Strings 4

1. 4 strings
2. 4 white, 0 black
3. 12
4. 30
5. Possible answer: There are 4 white beads for every 10 black, 14 in all. In a 42-bead string, there are $42 \div 14$, or 3, groups of 14 beads; 3×4, or 12, are white; 3×10, or 30, are black.

Bead Strings 5

1. 3 strings
2. 2 white, 3 black
3. 12
4. 16
5. Possible answer: A bead string with 28 beads will be twice as long as the one shown. It will have twice as many white beads, 12, and twice as many black beads, 16.

Bead Strings 6

1. 2 strings
2. Possible answer: Make a ratio table. There are not enough black and white beads to make 3 strings.
3. 10
4. 30
5. 20
6. Possible answer: For every 2 striped beads, there are 4 black and 6 white beads. This makes 12 beads; $60 \div 12 = 5$. There are 5 groups of 12 beads in 60, so the string will have 5×2, or 10, striped beads; 5×4, or 20, black beads; and 5×6, or 30, white beads.

Better Buy

Goals

- Compare items by identifying the relationship between quantity and cost.
- Generate equivalent ratios.
- Identify unit costs.

Notes

Prior to doing this problem set, you may want to review basic multiplication and division facts with students.

Solutions to all problems in this set appear on page 31.

Better Buy 1

Questions to Ask

- How much do two Animal Antics stickers cost? (9¢)
- How much do three Super Sticky stickers cost? (12¢)
- How many sets of Super Sticky stickers can you get for 36¢? (3 sets; $36 \div 12 = 3$)
- How many stickers do you get? (Each set has 3 stickers, so you would get 3×3, or 9, stickers.)

Solutions

1. 6

2. 24¢

3. Super Sticky

4. Possible answer: Compare the unit cost—The cost of one Super Sticky sticker is 12¢ ÷ 3, or 4¢. The cost of one Animal Antics sticker is 9¢ ÷ 2, or 4.5¢.

Better Buy 1

Carrie's Corner Store sells two kinds of stickers.

Animal Antics

2 for 9¢

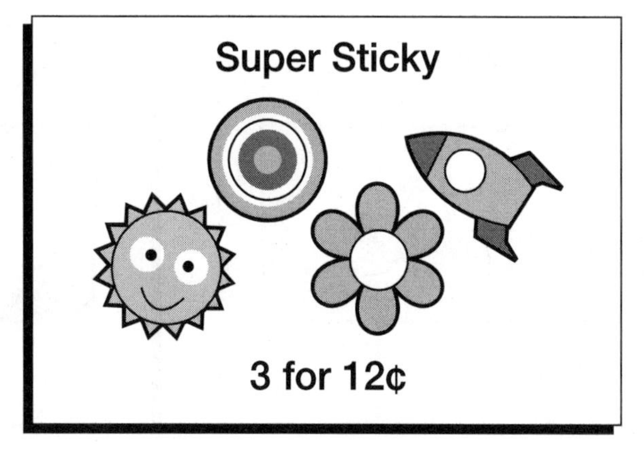

Super Sticky

3 for 12¢

1. How many Animal Antics stickers can you buy for 27¢? ____________
2. How much do six Super Sticky stickers cost? ____________
3. Which kind of sticker is the better buy? ____________
4. Tell how you know. ____________

____

____

Name ______________________________

Better Buy 2

Gau's Store sells two kinds of dried-fruit snacks.

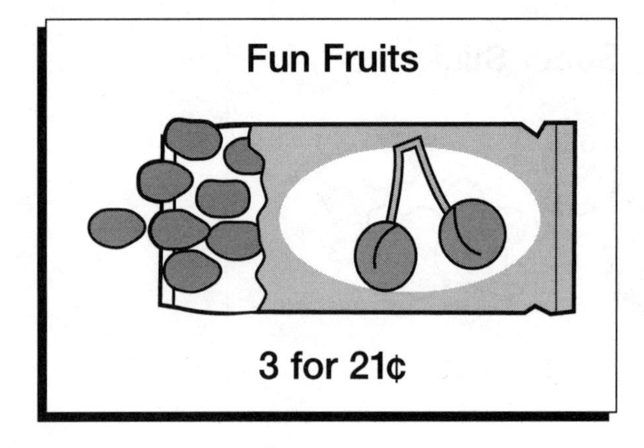

1. How many Fun Fruits can you buy for 63¢? ____________
2. How much do 12 Fruit Bars cost? ____________
3. Which snack is the better buy? ____________
4. Tell how you know. ____________

____

____

Name ____________________

Better Buy 3

Selina's Snacks sells two kinds of candy packs.

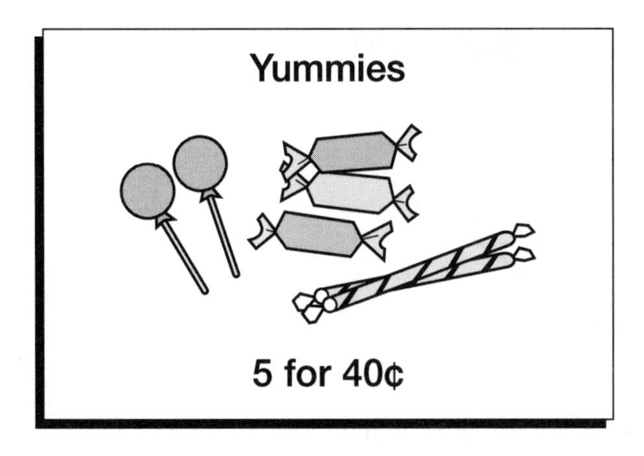

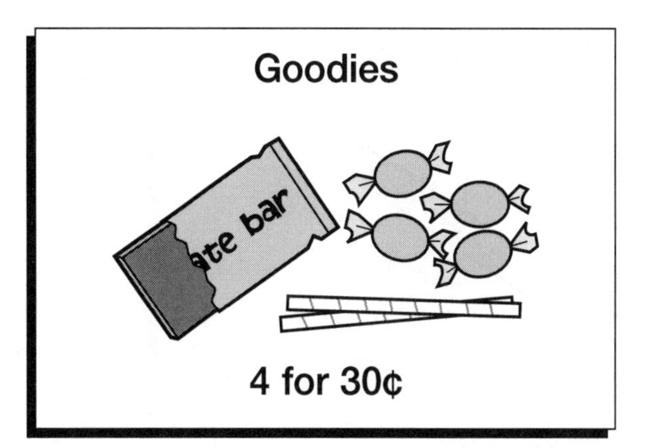

1. How many Goodies can you buy for 90¢? ____________
2. How much do 10 Yummies cost? ____________
3. Which candy pack is the better buy? ____________
4. Tell how you know. ____________

____

____

Name ______________________

Better Buy 4

School Supplies and The Pencil Store both sell pencils.

School Supplies

8 pencils for 72¢

The Pencil Store

10 pencils for 85¢

1. Which store has the better buy? ______________________
2. Explain how you know. ______________________
3. If you want to buy 40 pencils, to which store will you go? ______________________
4. Explain your answer. ______________________

Better Buy 5

Pen Colors and Ink It are stores that sell pens.

Pen Colors

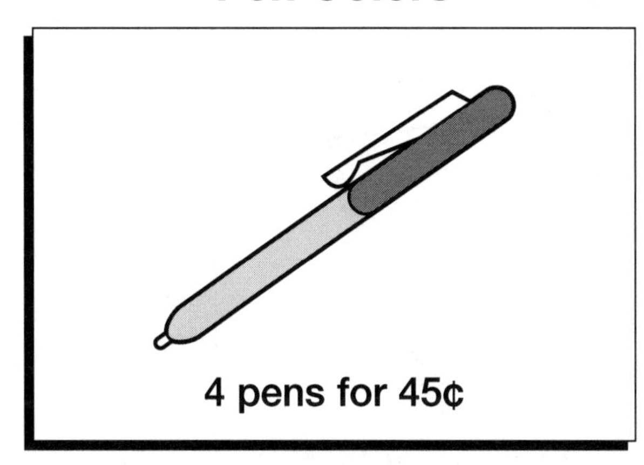

4 pens for 45¢

Ink It

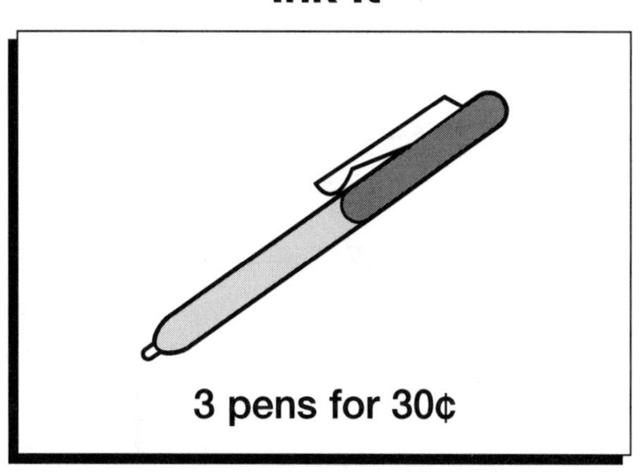

3 pens for 30¢

1. Which store has the better buy? ______________________
2. How do you know? ______________________

____

____

3. If you had 90¢ to spend on pens, what is the greatest number of pens you could buy? ____________
4. Where would you buy them? ______________________

Name ____________________

Better Buy 6

Paper Place and Paper House both sell pads of paper.

Paper Place

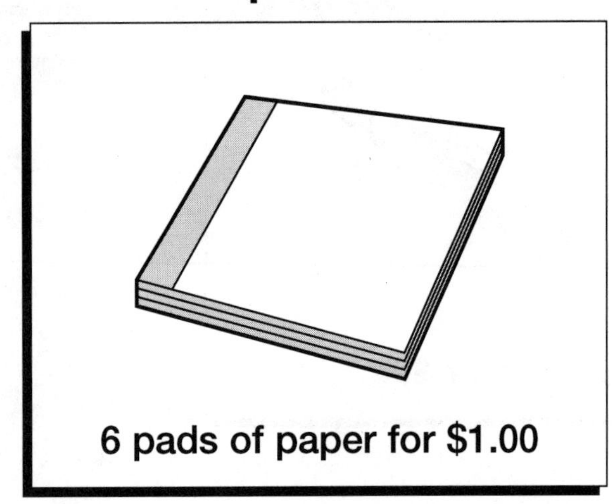

6 pads of paper for $1.00

Paper House

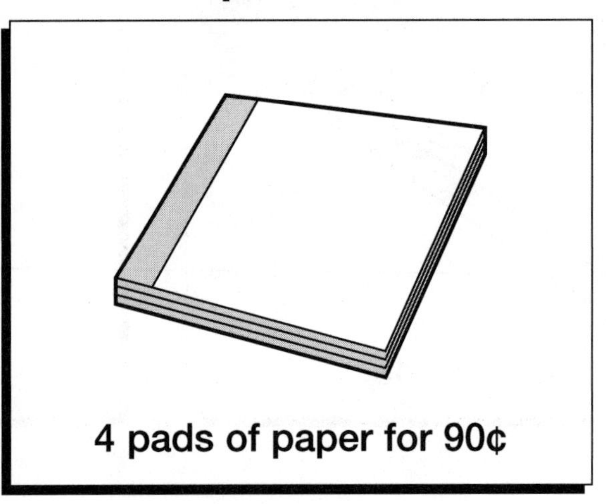

4 pads of paper for 90¢

1. Which store has the better buy? ____________________
2. Tell two ways to find the answer. ____________________

____

____

____

Solutions

Better Buy 1

1. 6
2. 24¢
3. Super Sticky
4. Possible answer: Compare the unit cost—The cost of one Super Sticky sticker is 12¢ ÷ 3, or 4¢. The cost of one Animal Antics sticker is 9¢ ÷ 2, or 4.5¢.

Better Buy 2

1. 9
2. 75¢
3. Fruit Bars
4. Possible answer: Find unit costs and compare—One Fun Fruit costs 21¢ ÷ 3, or 7¢. One Fruit Bar costs 25¢ ÷ 4, or a little more than 6¢.

Better Buy 3

1. 12
2. 80¢
3. Goodies
4. Possible answer: Find unit costs and compare—One Yummie costs 40¢ ÷ 5, or 8¢. One Goodie costs 30¢ ÷ 4, or 7.5¢.

Better Buy 4

1. The Pencil Store
2. Possible answer: Compare unit costs of the pencils—Eight pencils for 72¢ is 72¢ ÷ 8, or 9¢, per pencil at School Supplies. Ten pencils for 85¢ is 85¢ ÷ 10, or 8.5¢, per pencil at The Pencil Store.
3. The Pencil Store
4. Possible answer: At School Supplies, you can buy 40 pencils for 5 × 72¢, or $3.60. At The Pencil Store, you can buy 40 pencils for 4 × 85¢, or $3.40.

Better Buy 5

1. Ink It
2. Possible answer: Compare unit costs—At Pen Colors, one pen costs 45¢ ÷ 4, or a little more than 11¢. At Ink It, one pen costs 30¢ ÷ 3, or 10¢.
3. 9
4. Ink It

Better Buy 6

1. Paper Place
2. Possible answers: Compare the costs for 12 pads—At Paper Place, 12 pads cost 2 × $1, or $2. At Paper House, 12 pads cost 3 × 90¢, or $2.70. Twelve pads cost less at Paper Place. Compare unit costs—At Paper Place, $1.00 ÷ 6 is about 17¢ per pad. At Paper House, 90¢ ÷ 4 is 22.5¢ per pad.

Creature Trades

Goals

- Recognize trading as a proportional situation.
- Generate equivalent ratios.
- Construct ratio tables to solve problems.

Notes

To facilitate understanding of the proportional nature of the trading process, you might supply students with different color chips to represent the items to be traded and have them model the trades.

Solutions to all problems in this set appear on page 39.

Creature Trades 1

Questions to Ask

- How many pods will you get for two oogles? (four)
- How many oogles will you get for six pods? (three)
- If you know the number of oogles, how can you find the number of pods? (multiply the number of oogles by 2)
- If you know the number of pods, how can you find the number of oogles? (divide the number of pods by 2)

Solutions

1. 40 pods
2. 12 oogles
3. 31 pods
4. Possible answer: You can trade 13 oogles for 26 pods. This is fewer than 31 pods, so 31 pods is more than 13 oogles.

Name ______________________________

Creature Trades 1

You can trade:

1 oogle for 2 pods

or

2 pods for 1 oogle

1. If you trade 20 oogles for pods, how many pods will you get? __________
2. If you trade 24 pods for oogles, how many oogles will you get? __________
3. Which is more, 13 oogles or 31 pods? __________
4. Tell how you know. __________

____

____

____

Creature Trades 2

You can trade:

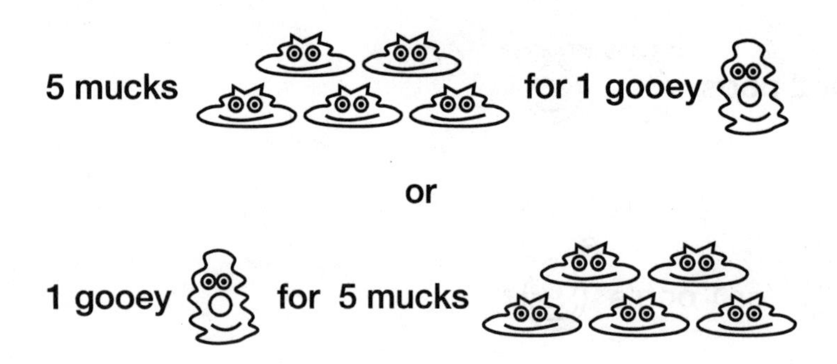

1. If you trade 10 mucks for gooeys, how many gooeys will you get? ___
2. If you trade 8 gooeys for mucks, how many mucks will you get? ___
3. Which is more, 15 mucks or 4 gooeys? ___
4. Tell how you know. ___

___

___

___

Creature Trades 3

You can trade:

2 blobs for 4 slimes

or

4 slimes for 2 blobs

1. How many slimes can you get for 20 blobs? ______

2. How many blobs can you get for 16 slimes? ______

3. Jack traded 60 slimes for 40 blobs.

   Is this a fair trade? ______

4. Explain how you know. ______

______

______

______

Creature Trades 4

3 glums = 5 gloms

1. Jan traded 9 glums for gloms. How many did she get? ________
2. How many gloms would you trade to get 24 glums? ________
3. Jon traded 20 gloms for 12 glums. Was it a fair trade? ________
4. Explain how you know. ____________________

____

____

____

Creature Trades 5

5 shmoos = 6 fuds

1. How many fuds would you trade to get 15 shmoos? ________
2. How many shmoos would you get for 24 fuds? ________
3. Is 30 shmoos for 40 fuds a fair trade? ________
4. Explain how you know. ____________________

____

____

____

Name ______________________________

Creature Trades 6

1 ping = 2 pongs = 4 pogos

1. How many pogos would you trade to get 16 pongs? __________

2. Would you get more pings or more pogos for 8 pongs? __________

3. How do you know? ______________________________

____

____

4. How many more pongs than pings would you get for 48 pogos? __________

5. Explain how you know. ______________________________

____

____

____

Solutions

Creature Trades 1

1. 40 pods
2. 12 oogles
3. 31 pods
4. Possible answer: You can trade 13 oogles for 26 pods. This is fewer than 31 pods, so 31 pods is more than 13 oogles.

Creature Trades 2

1. 2 gooeys
2. 40 mucks
3. 4 gooeys
4. Possible answer: You can trade 15 mucks for 3 gooeys, so 4 gooeys is more than 15 mucks.

Creature Trades 3

1. 40 slimes
2. 8 blobs
3. no
4. Possible answer: The number of slimes is twice the number of blobs. For 40 blobs, you would get 2×40, or 80, slimes.

Creature Trades 4

1. 15 gloms
2. 40 gloms
3. yes
4. Possible answer: 20 gloms is 4 sets of 5 gloms. Then 4 sets of 3 glums is 12 glums, so the trade was fair.

Creature Trades 5

1. 18 fuds
2. 20 shmoos
3. no
4. Possible answer: For 30 shmoos (6 sets of 5 schmoos), you would get 6×6, or 36, fuds. Since 40 is more than 36, the trade is not fair.

Creature Trades 6

1. 32 pogos
2. more pogos
3. Possible answer: The number of pings is half the number of pongs. For 8 pongs, you get 4 pings. The number of pogos is twice the number of pongs. For 8 pongs, you get 16 pogos.
4. 12 more pongs
5. Possible answer: For 48 pogos, you get $48 \div 2$, or 24, pongs. For 48 pogos, you get $48 \div 4$, or 12, pings. You get $24 - 12$, or 12, more pongs than pings.

Balance Beam

Goals

- Recognize that a balanced beam represents equality.
- Compute the moment of an object.
- Write an equation to represent a balanced beam.

Notes

Encourage students to figure out the total of the moments (the product of an object's weight and its distance from the center of the beam) on each side of the balance beam and then compute the difference. As students become more experienced, they may recognize that equivalent weights on each side can be crossed off to simplify the problem.

Solutions to all problems in this set appear on page 47.

Balance Beam 1

Questions to Ask

- Each circle represents a mass. How much does each mass weigh? (1 ounce)
- How many masses are hanging from the 1? (3)
- What is the moment of the masses hanging from the 1? ($3 \times 1 = 3$)
- What is the moment of the mass hanging from the 2? ($1 \times 2 = 2$)

Solutions

1. 5

2. Possible answers: The total moment of the left arm is $1 \times 2 + 3 \times 1$, or 5. To make the beam balance, hang one mass from the 5 on the right arm.

Write an equation.

$1 \times 2 + 3 \times 1 = 1 \times M$

$2 + 3 = M$

$5 = M$

Name ______________________________

Balance Beam 1

Each mass (○) weighs 1 ounce.

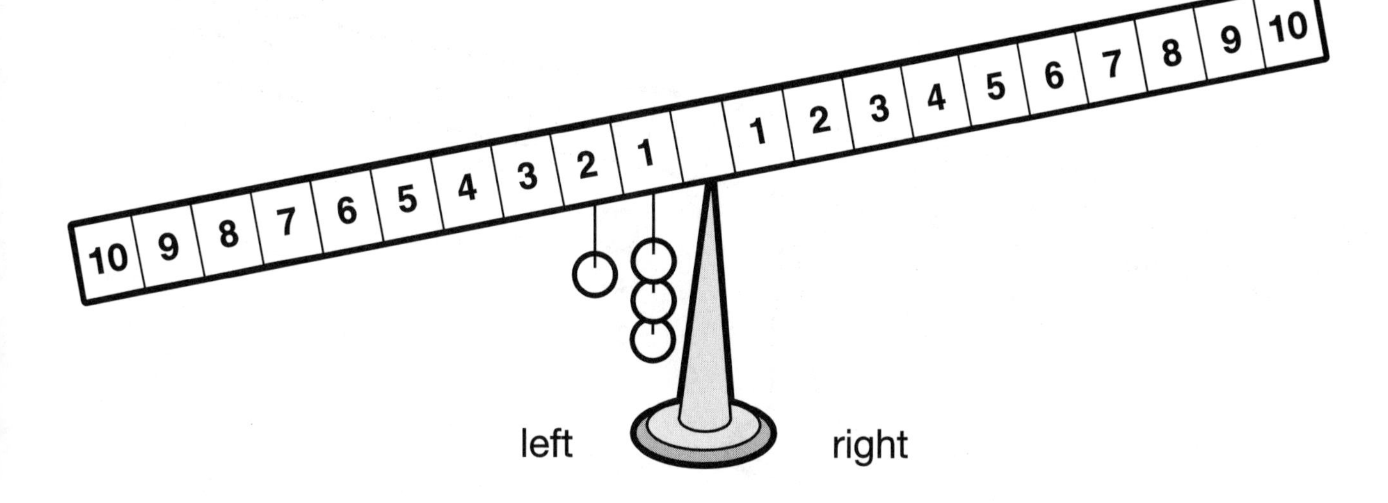

Put one mass on the right side to make the beam balance.

1. The mass will hang from number ________.
2. Explain how you know. ______________________________

____

____

Name ______________________________

Balance Beam 2

Each mass (○) weighs 1 ounce.

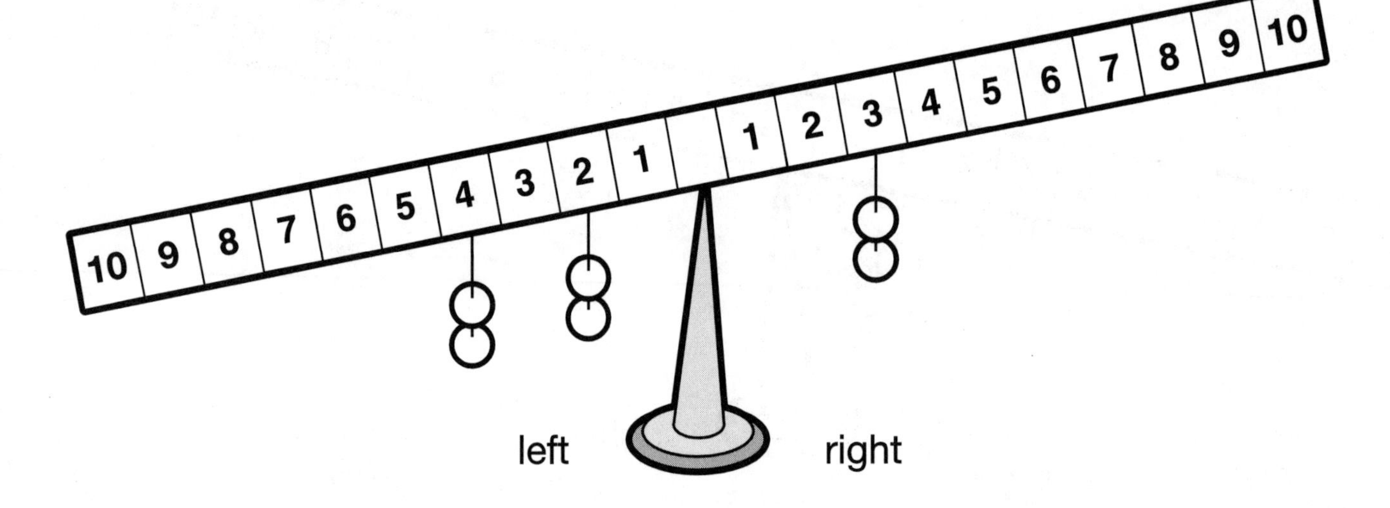

Put one mass on the right side to make the beam balance.

1. The mass will hang from number __________.
2. Explain how you know. ______________________________

____

____

Name ____

Balance Beam 3

Each mass (○) weighs 1 ounce.

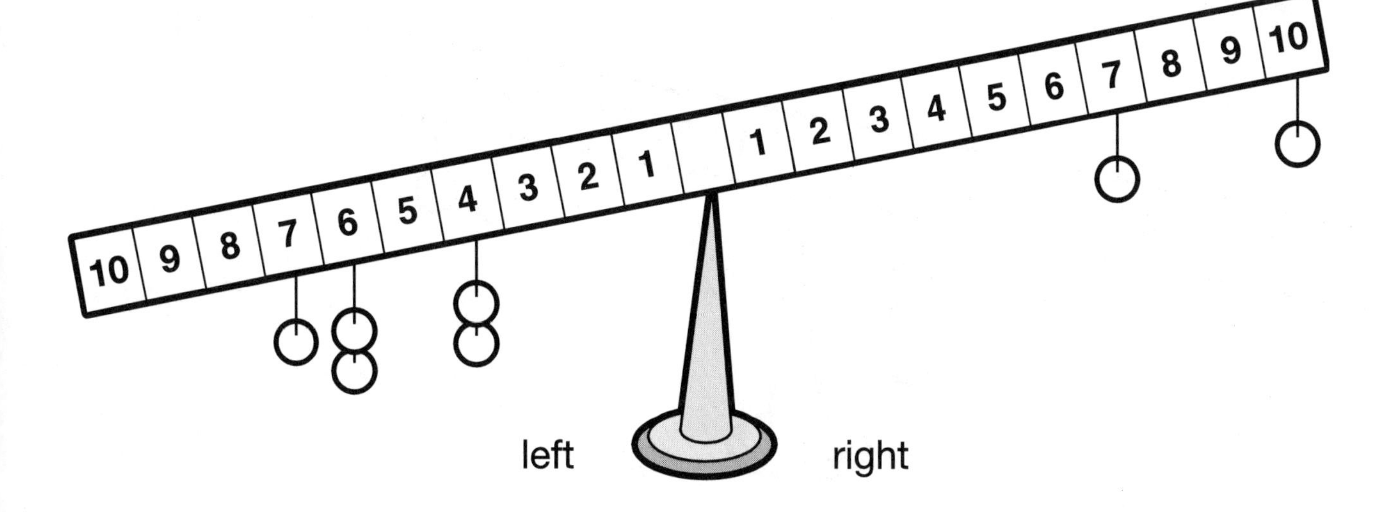

Put one mass on the right side to make the beam balance.

1. The mass will hang from number ________.

2. Explain how you know. ______________________________

____

____

Balance Beam 4

Each mass (o) weighs 1 ounce.

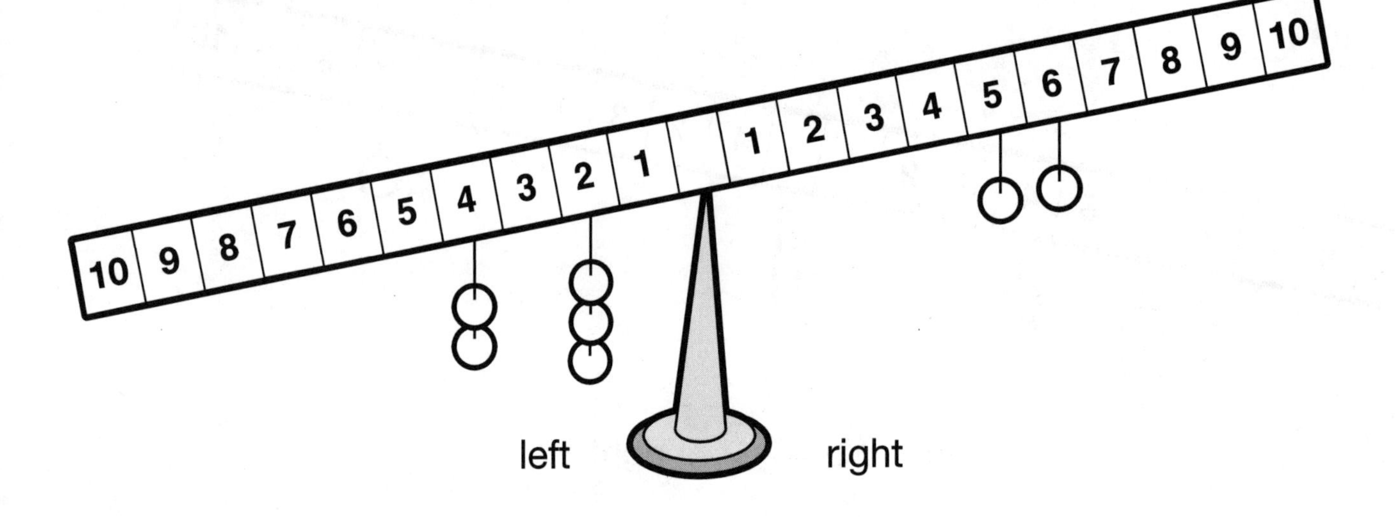

Put one mass on the right side to make the beam balance.

1. The mass will hang from number ________.

2. Explain how you know. ______________________________

____

____

Balance Beam 5

Each mass (o) weighs 1 ounce.

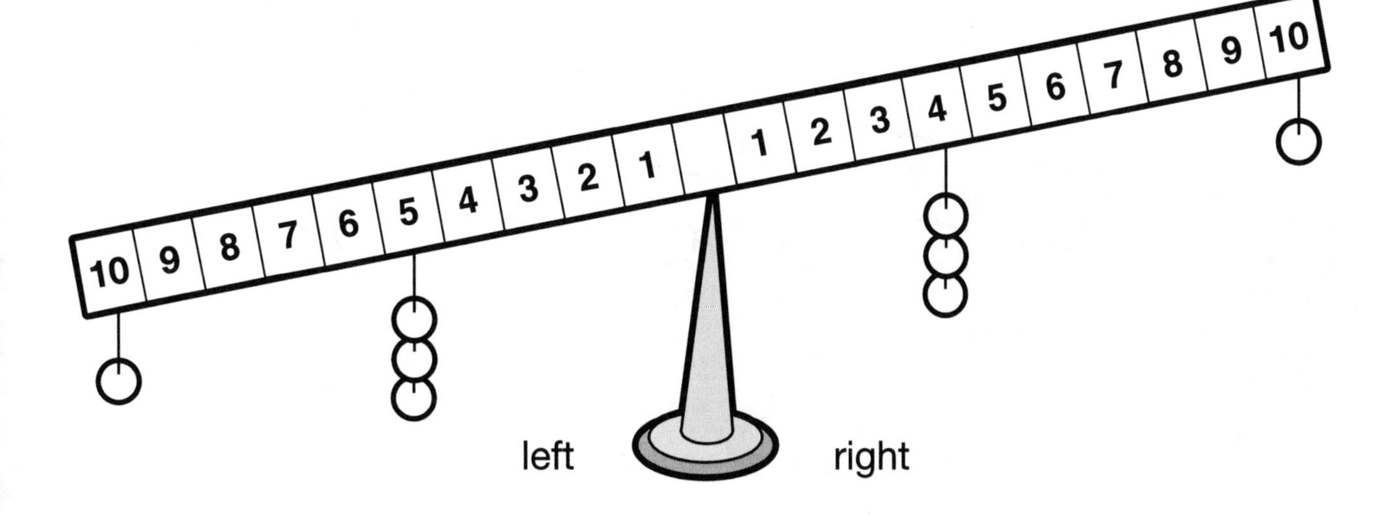

Put one mass on the right side to make the beam balance.

1. The mass will hang from number ________.

2. Explain how you know. ____________________

____

____

Balance Beam 6

Each mass (○) weighs 1 ounce.

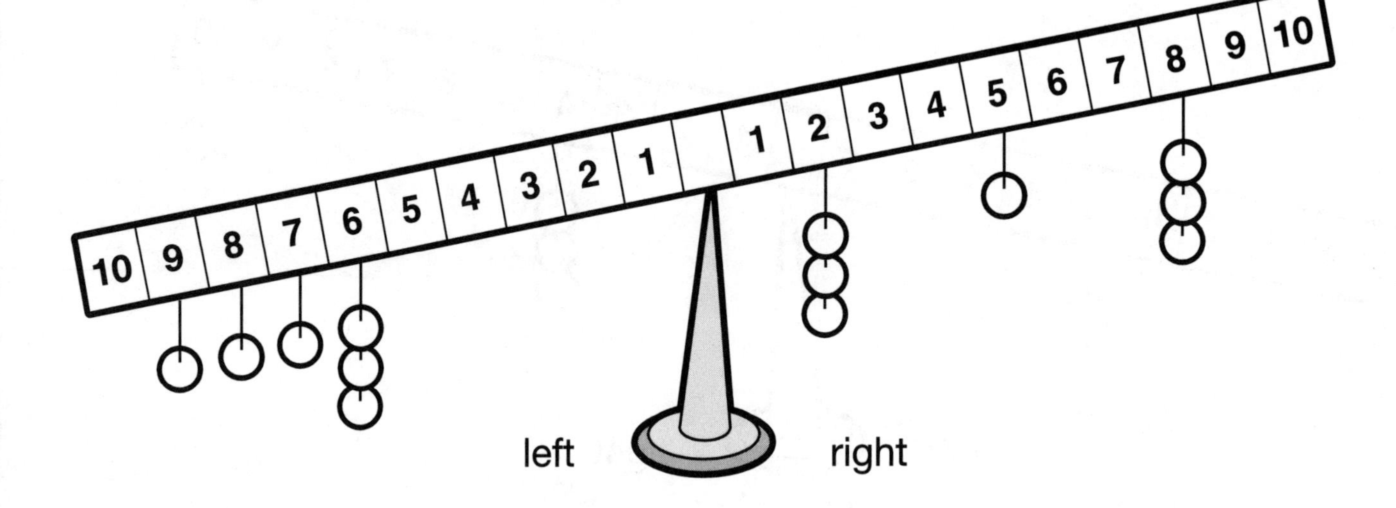

Put one mass on the right side to make the beam balance.

1. The mass will hang from number __________.

2. Explain how you know. ______________________________

____

____

Solutions

Balance Beam 1

1. 5
2. Possible answers: The total moment of the left arm is $1 \times 2 + 3 \times 1$, or 5. To make the beam balance, hang one mass from the 5 on the right arm.

   Write an equation.
   $1 \times 2 + 3 \times 1 = 1 \times M$
   $2 + 3 = M$
   $5 = M$

Balance Beam 2

1. 6
2. Possible answers: Remove one mass from the 4 and one from the 2 on the left and both masses from the 3 on the right. That leaves one mass hanging from the 4 and one from the 2 on the left. To balance them, hang one mass from the 6 on the right.

   Write an equation.
   $2 \times 4 + 2 \times 2 = 2 \times 3 + 1 \times M$
   $8 + 4 = 6 + M$
   $12 = 6 + M$
   $6 = M$

Balance Beam 3

1. 10
2. Possible answers: Remove the 7s from each side. Then remove a mass from the 4 and the 6 on the left and the 10 on the right. That leaves one 6 and one 4 on the left. To balance them, hang a mass from 10 on the right arm.

   Write an equation.
   $1 \times 7 + 2 \times 6 + 2 \times 4 =$
   $1 \times 7 + 1 \times 10 + 1 \times M$
   $7 + 12 + 8 = 7 + 10 + M$
   $27 = 17 + M$
   $10 = M$

Balance Beam 4

1. 3
2. Possible answer:
   Write an equation.
   $2 \times 4 + 3 \times 2 =$
   $1 \times 5 + 1 \times 6 + 1 \times M$
   $8 + 6 = 5 + 6 + M$
   $14 = 11 + M$
   $3 = M$

Balance Beam 5

1. 3
2. Possible answer:
   Write an equation.
   $1 \times 10 + 3 \times 5 =$
   $3 \times 4 + 1 \times 10 + 1 \times M$
   $10 + 15 = 12 + 10 + M$
   $25 = 22 + M$
   $3 = M$

Balance Beam 6

1. 7
2. Possible answer:
   Write an equation.
   $1 \times 9 + 1 \times 8 + 1 \times 7 + 3 \times 6 =$
   $3 \times 2 + 1 \times 5 + 3 \times 8 + 1 \times M$
   $9 + 8 + 7 + 18 = 6 + 5 + 24 + M$
   $42 = 35 + M$
   $7 = M$

Pan Balances

Goals

- Deduce mass relationships from visual clues.
- Recognize that a level balance represents equality.
- Identify collections of objects having equal mass.
- Use substitution as a method for solving equations.

Notes

Encourage students to simplify each pan balance by "removing" (crossing off) equivalent weights on each side. Have students look to see which pan holds a subset of the set of blocks in another pan. Substituting one set of blocks for another may facilitate the solution process.

Solutions to all problems in this set appear on page 55.

Pan Balances 1

Questions to Ask

- Which balances are level? (A and B)
- Look at B. Would the balance be level if you took the cylinder off the left pan and a cylinder off the right pan? (yes) Why? (because the same mass came off each side)
- Look at A. If you added one cube to the left pan and one cube to the right pan, would the balance still be level? (Yes, when you add the same mass to each side, a balance stays level.)

Solutions

1. cube
2. Possible answer: On B, remove one cylinder from each pan. That leaves one cube balancing two cylinders. Using A, substitute three spheres for each cylinder on the right pan of B. That leaves one cube balancing six spheres. On C, put a cube on the left pan to balance the six spheres on the right pan.
3. 9, 18
4. 2, 12

Name ______________________________

Pan Balances 1

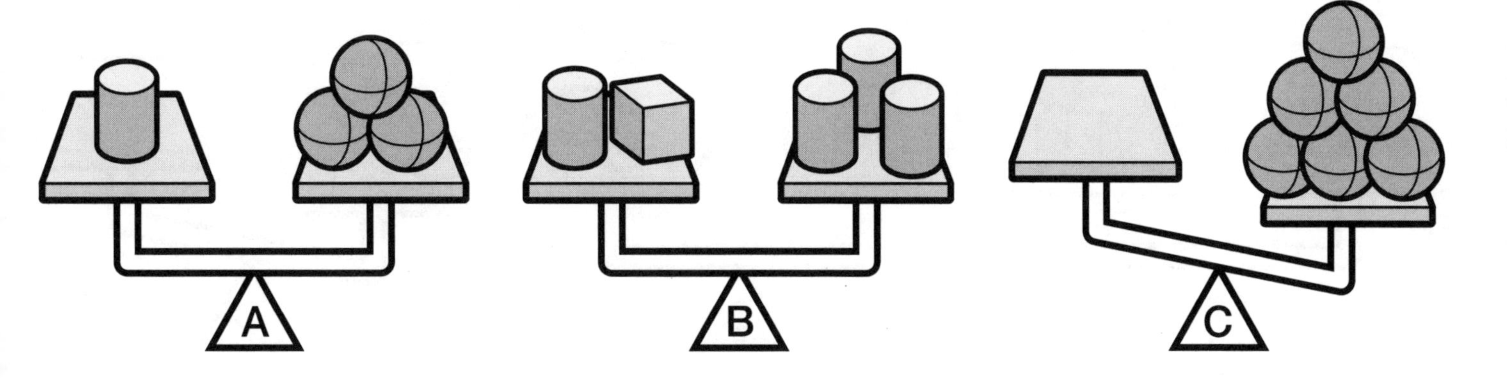

1. Which block will balance C? ______________________

2. Tell how you chose the block. ______________________

____

____

3. If the sphere weighs 3 pounds, then

   the cylinder weighs __________ pounds, and

   the cube weighs __________ pounds.

4. If the cylinder weighs 6 pounds, then

   the sphere weighs __________ pounds, and

   the cube weighs __________ pounds.

Pan Balances 2

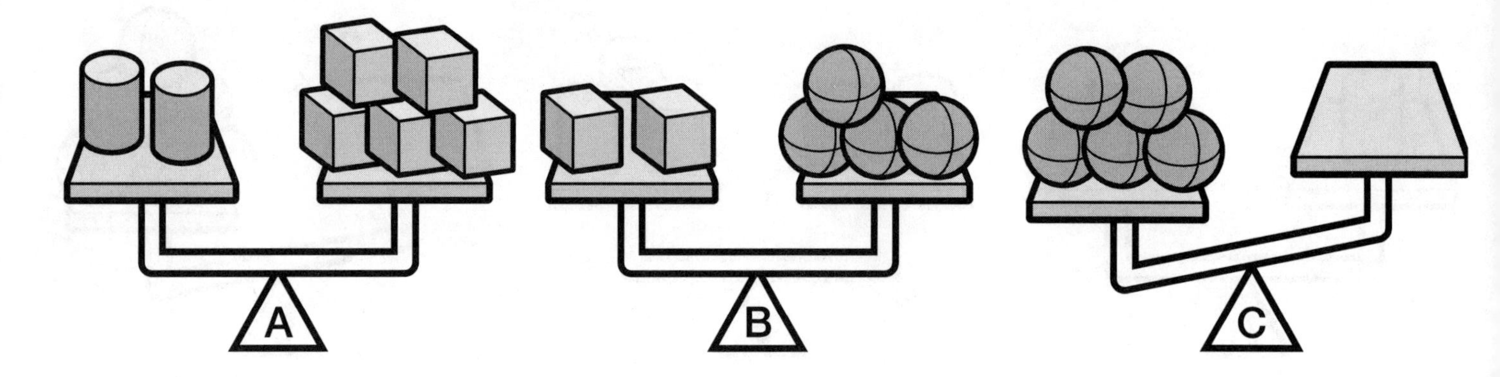

1. Which block will balance C? ______________________

2. Tell how you chose the block. ______________________

____

____

3. If the sphere weighs 2 pounds, then

   the cube weighs __________ pounds, and

   the cylinder weighs __________ pounds.

4. If the cube weighs 2 pounds, then

   the cylinder weighs __________ pounds, and

   the sphere weighs __________ pound.

Name ________________________________

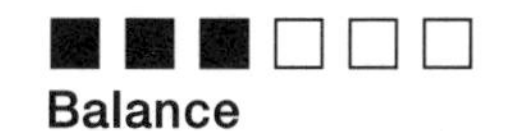

Pan Balances 3

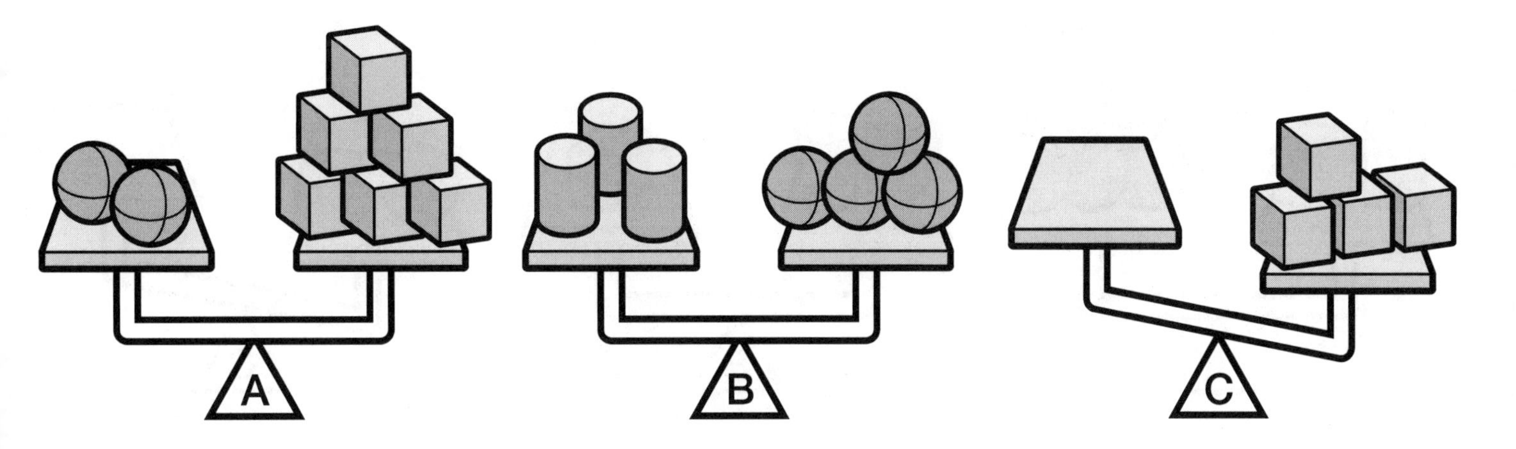

1. Which block will balance C? ____________________

2. Tell how you chose the block. ____________________

____

____

3. If the cube weighs 2 pounds, then

   the sphere weighs __________ pounds, and

   the cylinder weighs __________ pounds.

4. If the sphere weighs 9 pounds, then

   the cylinder weighs __________ pounds, and

   the cube weighs __________ pounds.

Pan Balances 4

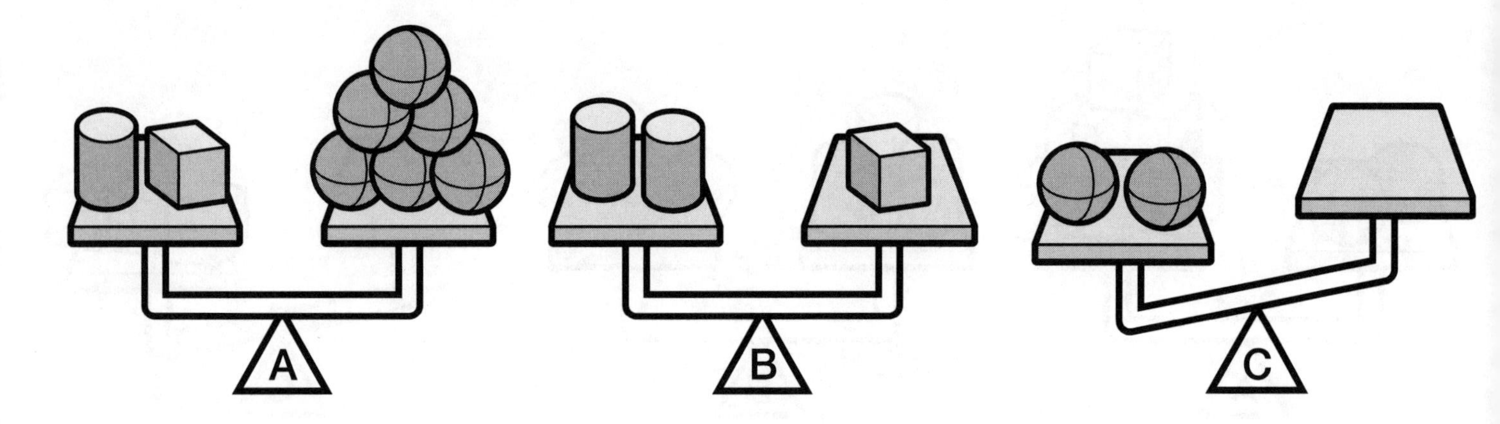

1. Which block will balance C? ______________________

2. Tell how you chose the block. ______________________

____

____

3. If the cylinder weighs 4 pounds, then

   the cube weighs ________ pounds, and

   the sphere weighs ________ pounds.

4. If the cube weighs 12 pounds, then

   the cylinder weighs ________ pounds, and

   the sphere weighs ________ pounds.

Name ______________________________

Pan Balances 5

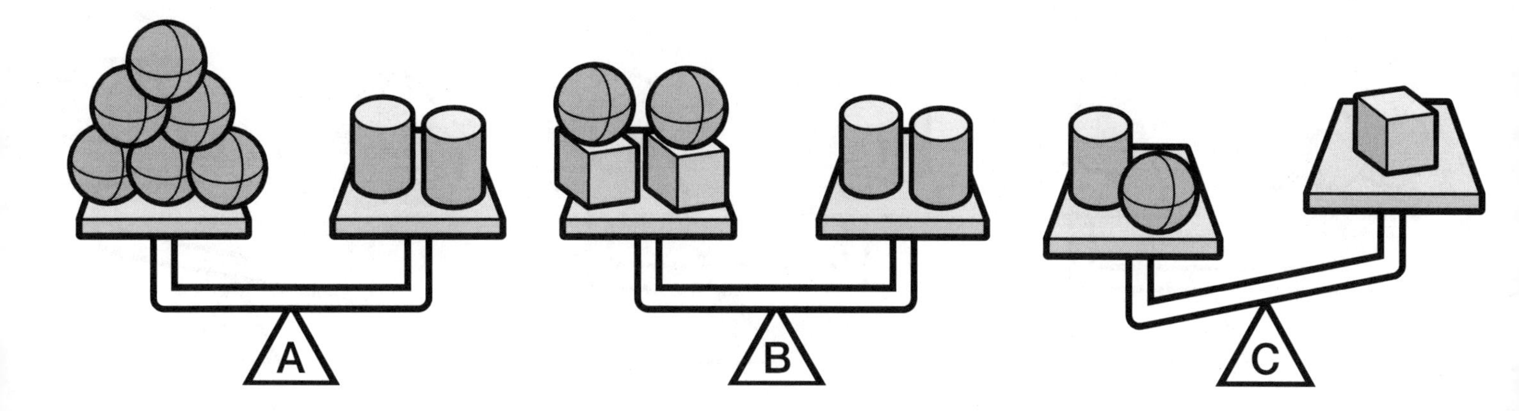

1. Which block will balance C? ______________________

2. Tell how you chose the block. ______________________

____

____

3. If the sphere weighs 2 pounds, then

   the cylinder weighs __________ pounds, and

   the cube weighs __________ pounds.

4. If the cylinder weighs 9 pounds, then

   the sphere weighs __________ pounds, and

   the cube weighs __________ pounds.

Pan Balances 6

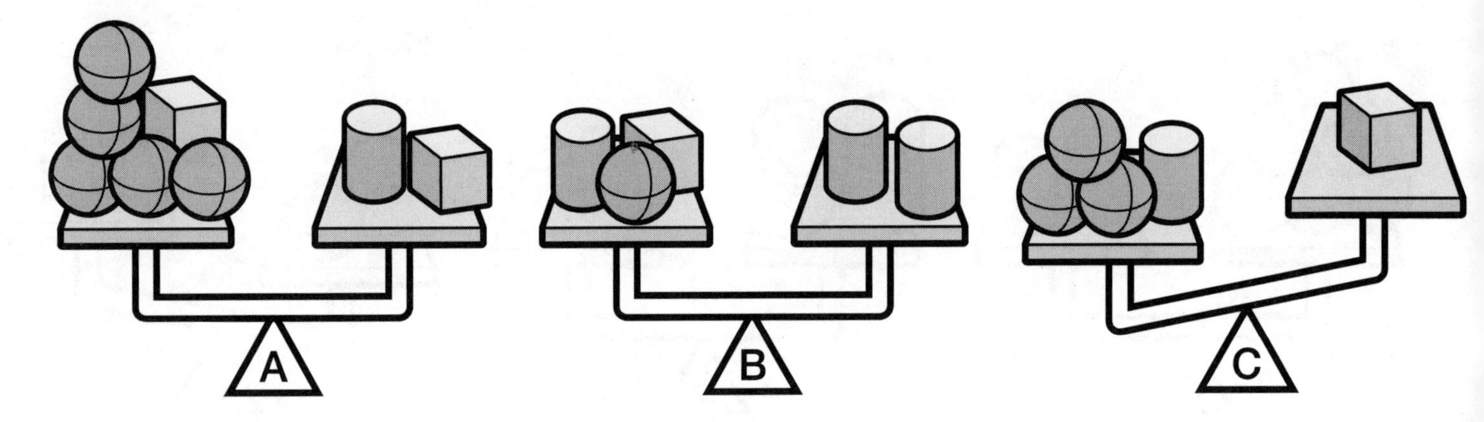

1. Which block will balance C? ____________________

2. Tell how you chose the block. ____________________

____

____

3. If the sphere weighs 2 pounds, then

the cylinder weighs __________ pounds, and

the cube weighs __________ pounds.

4. If the cylinder weighs 15 pounds, then

the sphere weighs __________ pounds, and

the cube weighs __________ pounds.

Solutions

Pan Balances 1

1. cube
2. Possible answer: On B, remove one cylinder from each pan. That leaves one cube balancing two cylinders. Using A, substitute three spheres for each cylinder on the right pan of B. That leaves one cube balancing six spheres. On C, put a cube on the left pan to balance the six spheres on the right pan.
3. 9, 18
4. 2, 12

Pan Balances 2

1. cylinder
2. Possible answer: From B, one cube balances two spheres. Substitute spheres for cubes on the right pan of A. That leaves two cylinders on the left balancing ten spheres on the right. One cylinder balances five spheres. Put one cylinder on the right pan of C to make the balance level.
3. 4, 10
4. 5, 1

Pan Balances 3

1. cylinder
2. Possible answer: From A, one sphere balances three cubes. Substitute three cubes for each sphere on B. That leaves three cylinders balancing 12 cubes, so one cylinder balances four cubes. Put a cylinder on the left pan of C to level the balance.
3. 6, 8
4. 12, 3

Pan Balances 4

1. cylinder
2. Possible answer: Using B, substitute two cylinders for the cube on the left pan of A. Then three cylinders balance six spheres, so one cylinder balances two spheres. The left pan of C has two spheres. Put one cylinder on the right pan to balance C.
3. 8, 2
4. 6, 3

Pan Balances 5

1. cube
2. Possible answer: Using A, substitute six spheres for two cylinders on B. Then remove two spheres from each pan of B. This leaves two cubes balancing four spheres, so one cube balances two spheres. On C, substitute three spheres for the cylinder on the left pan. Since four spheres balances two cubes, put another cube on the right pan to balance C.
3. 6, 4
4. 3, 6

Pan Balances 6

1. cube
2. Possible answer: Remove one cube from each side of A so five spheres balance one cylinder. On B, substitute five spheres for each cylinder. Then one cube has a mass equal to four spheres. On C, replace the cylinder with five spheres, so there are eight spheres on the left. The right pan has one cube, which has the same mass as four spheres. Put one cube on the right pan to balance C.
3. 10, 8
4. 3, 12

Weighing Blocks

Goals

- Identify values of blocks from relationships shown symbolically.
- Replace variables with numbers in systems of equations.
- Use substitution as a solution method.
- Make inferences.

Notes

In all problems in this set, one scale has only one type of block (variable). Encourage students to identify that scale first and then solve for the value of that block. Once they have solved for the weights of the blocks, have them replace each block on each scale with its value and check that the sum of the weights equals the weight shown by the scale.

Solutions to all problems in this set appear on page 63.

Weighing Blocks 1

Questions to Ask

- What blocks are on Scale C? (a cylinder and a cube)
- What is the total weight of the blocks on Scale B? (10 pounds)
- Which is heavier, a cube or a sphere? (a cube)
- How do you know? (Scales B and C each have a cylinder, so their difference in weight must come from the other block. Scale C is heavier by 1 pound, so a cube must be 1 pound heavier than a sphere.)

Solutions

1. sphere = 4 pounds
   cylinder = 6 pounds
   cube = 5 pounds
2. Possible answer: On Scale A, one sphere weighs $12 \div 3$, or 4, pounds. On Scale B, since the sphere weighs 4 pounds, the cylinder weighs $10 - 4$, or 6, pounds. On Scale C, since the cylinder weighs 6 pounds, the cube weighs $11 - 6$, or 5, pounds.

Name ____________________

Weighing Blocks 1

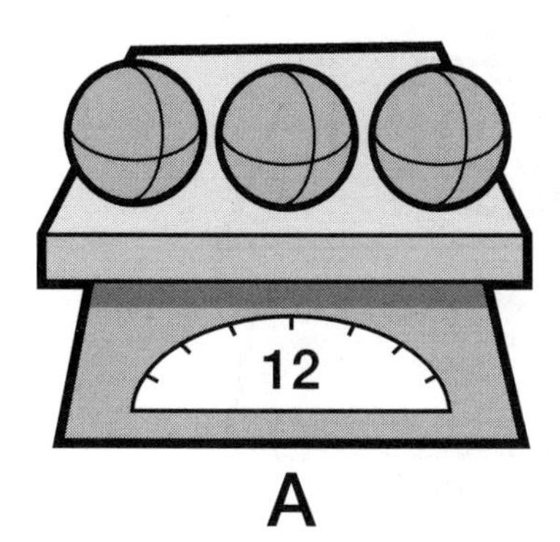

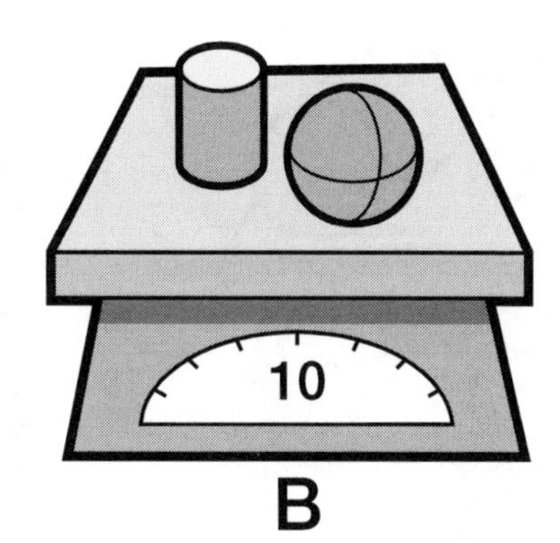

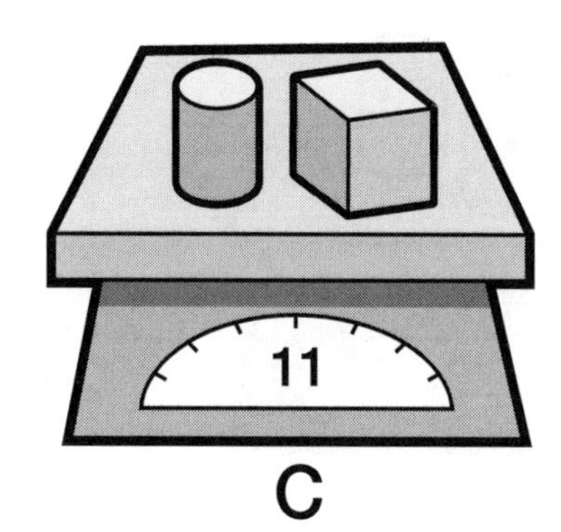

1. Find the weight of each block.

sphere = __________ pounds

cylinder = __________ pounds

cube = __________ pounds

2. Tell how you figured it out. ____________________

____

____

____

Name ______________________________

Weighing Blocks 2

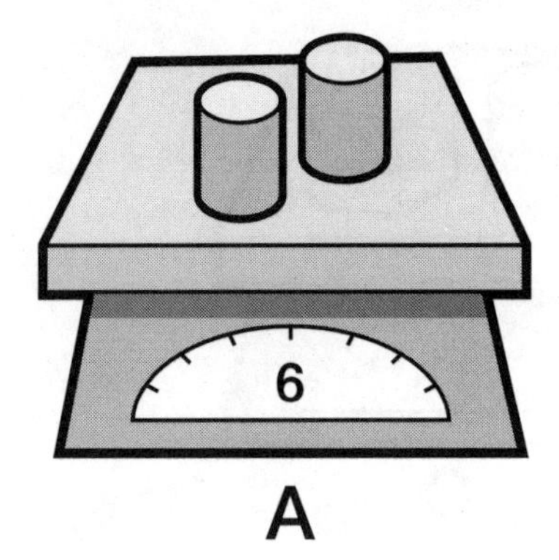

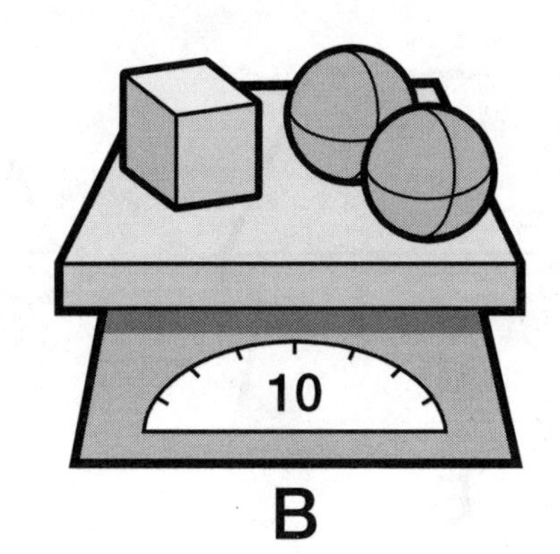

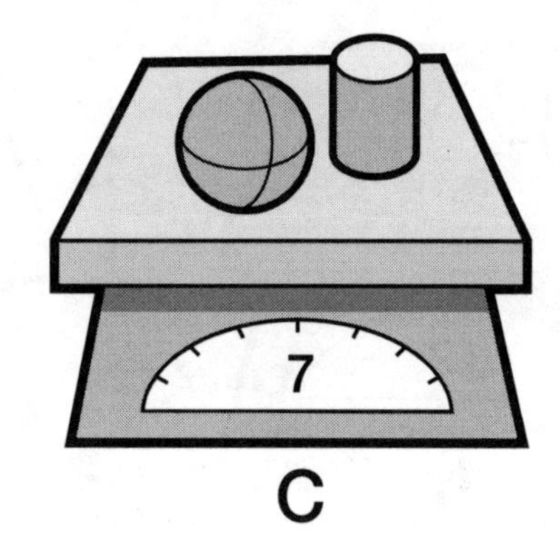

1. Find the weight of each block.

   sphere = __________ pounds

   cylinder = __________ pounds

   cube = __________ pounds

2. Tell how you figured it out. ______________________________

   ____

   ____

   ____

Name ______________________

■■■■□□ Variable

Weighing Blocks 3

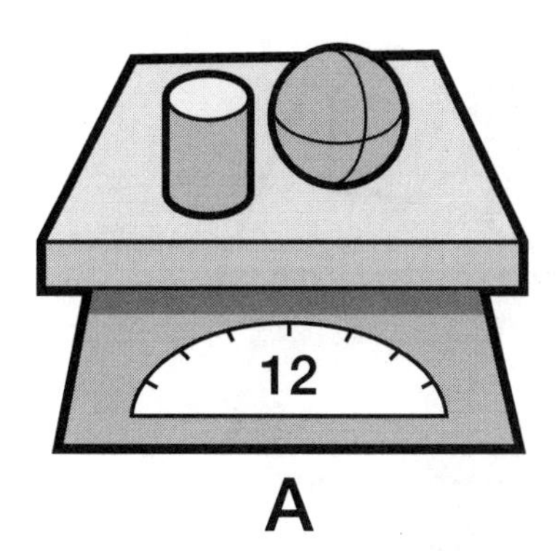

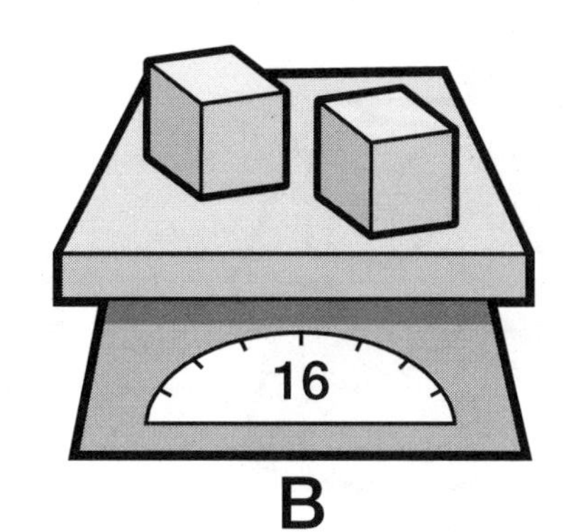

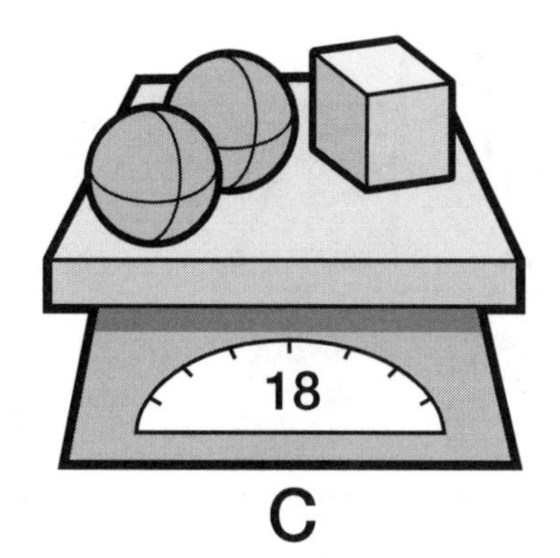

1. Find the weight of each block.

   sphere = __________ pounds

   cylinder = __________ pounds

   cube = __________ pounds

2. Tell how you figured it out. ______________________________

   ____

   ____

   ____

Weighing Blocks 4

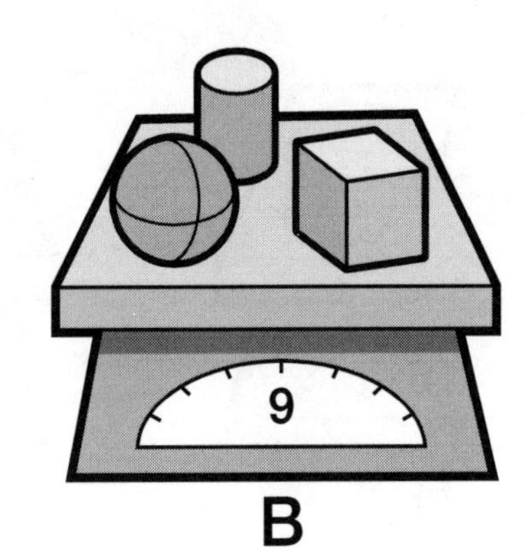

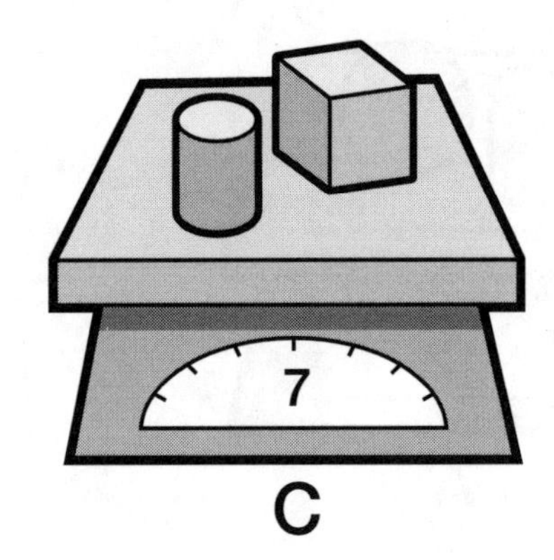

1. Find the weight of each block.

   sphere = __________ pounds

   cylinder = __________ pounds

   cube = __________ pounds

2. Tell how you figured it out. ______________________________

   ____

   ____

   ____

Weighing Blocks 5

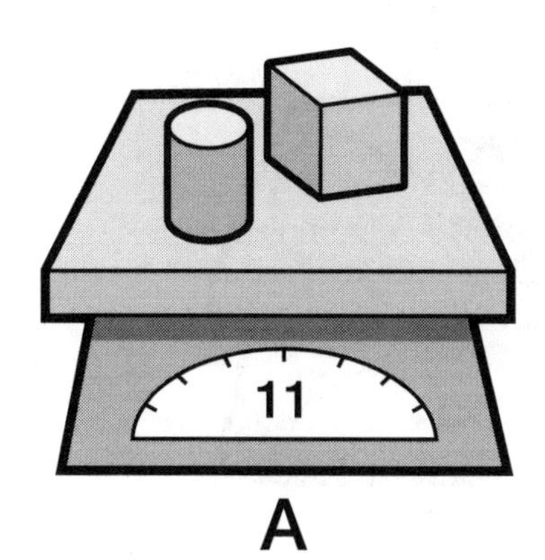

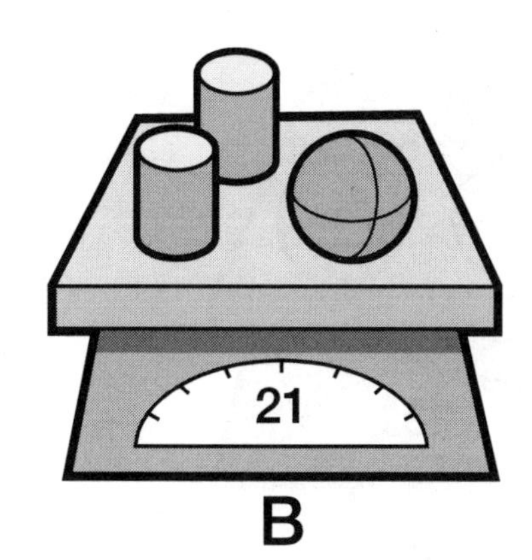

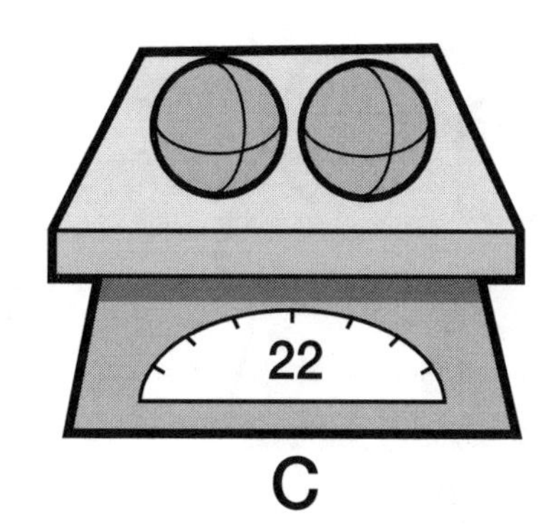

1. Find the weight of each block.

   sphere = __________ pounds

   cylinder = __________ pounds

   cube = __________ pounds

2. Tell how you figured it out. ______________________________

____

____

____

Name ______________________

Variable

Weighing Blocks 6

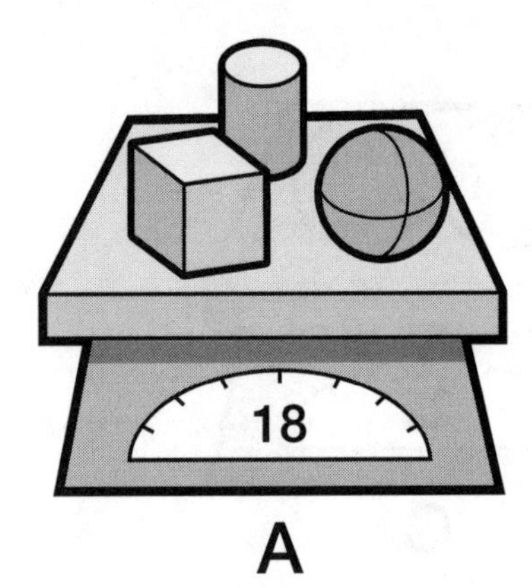

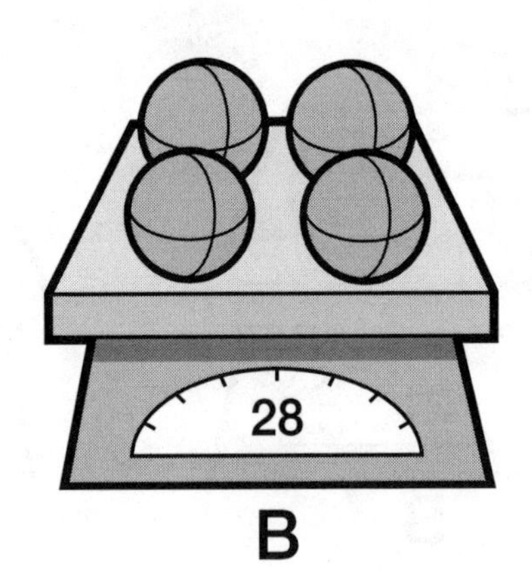

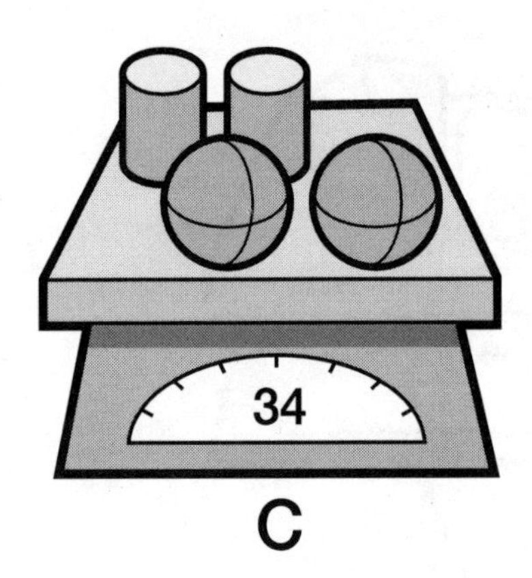

1. Find the weight of each block.

   sphere = __________ pounds

   cylinder = __________ pounds

   cube = __________ pound

2. Tell how you figured it out. ______________________________

   ____

   ____

   ____

Solutions

Weighing Blocks 1

1. sphere = 4 pounds
   cylinder = 6 pounds
   cube = 5 pounds
2. Possible answer: On Scale A, one sphere weighs 12 ÷ 3, or 4, pounds. On Scale B, since the sphere weighs 4 pounds, the cylinder weighs 10 − 4, or 6, pounds. On Scale C, since the cylinder weighs 6 pounds, the cube weighs 11 − 6, or 5, pounds.

Weighing Blocks 2

1. sphere = 4 pounds
   cylinder = 3 pounds
   cube = 2 pounds
2. Possible answer: On Scale A, two cylinders weigh 6 pounds, so one cylinder weighs 6 ÷ 2, or 3, pounds. On Scale C, since the cylinder weighs 3 pounds, the sphere weighs 7 − 3, or 4, pounds. On Scale B, since each sphere weighs 4 pounds, the cube weighs 10 − 8, or 2, pounds.

Weighing Blocks 3

1. sphere = 5 pounds
   cylinder = 7 pounds
   cube = 8 pounds
2. Possible answer: On Scale B, two cubes weigh 16 pounds, so each cube weighs 8 pounds. On Scale C, since the cube weighs 8 pounds, the two spheres weigh 18 − 8, or 10, pounds; each sphere weighs 5 pounds. On Scale A, since the sphere weighs 5 pounds, the cylinder weighs 12 − 5, or 7, pounds.

Weighing Blocks 4

1. sphere = 2 pounds
   cylinder = 3 pounds
   cube = 4 pounds
2. Possible answer: The weight of the three cylinders on Scale A is 9 pounds, so each cylinder weighs 3 pounds. If the cylinder on Scale C weighs 3 pounds, then the cube weighs 7 − 3, or 4, pounds. If the cylinder and the cube on Scale B together weigh 7 pounds, then the sphere weighs 9 − 7, or 2, pounds.

Weighing Blocks 5

1. sphere = 11 pounds
   cylinder = 5 pounds
   cube = 6 pounds
2. Possible answer: If the two spheres on Scale C weigh 22 pounds, each sphere weighs 11 pounds. If the sphere on Scale B weighs 11 pounds, the cylinders weigh 21 − 11, or 10, pounds; each cylinder weighs 5 pounds. If the cylinder on Scale A weighs 5 pounds, then the cube weighs 6 pounds.

Weighing Blocks 6

1. sphere = 7 pounds
   cylinder = 10 pounds
   cube = 1 pound
2. Possible answer: If the spheres on Scale B weigh 28 pounds, each sphere weighs 28 ÷ 4, or 7, pounds. If each sphere on Scale C weighs 7 pounds, the two cylinders weigh 34 − 14, or 20, pounds; each cylinder weighs 10 pounds. If the cylinder and the sphere on Scale A weigh 17 pounds, then the cube weighs 18 − 17, or 1, pound.

Missing Numbers

Goals

- Replace variables with numbers in systems of equations.
- Identify relationships among variables.
- Use substitution as a method for solving systems of equations.

Notes

The problems in this set are different representations of the problems in the preceding set. Call students' attention to the relationship between these two representations. Encourage them to approach the problems in the same way.

Solutions to all problems in this set appear on page 71.

Missing Numbers 1

Questions to Ask

- How are the equations different? (There is one more circle in the first equation; the sums are different.)
- How are the equations alike? (Both equations have a square and a circle.)
- In the second equation, if Square is 2, what is the value of Circle? ($11 - 2$, or 9)
- What pairs of whole numbers can you put in Square and Circle to make the second equation true? (0, 11; 1, 10; 2, 9; 3, 8; 4, 7; 5, 6; 6, 5; 7, 4; 8, 3; 9, 2; 10, 1; 11, 0)

Solutions

1. 6
2. 5
3. Possible answer: Both equations have a square and a circle. Replace the square and one circle in the first equation with 11. The new equation is $11 +$ Circle $= 16$. Circle is 5. In the second equation, if Circle is 5, Square is $11 - 5$, or 6.

Name ______________________

Missing Numbers 1

Same shapes are same numbers.
Different shapes are different numbers.

$\square + \bigcirc + \bigcirc = 16$

$\square + \bigcirc = 11$

1. What number is Square? __________

2. What number is Circle? __________

3. Tell how you found the numbers. ______________________

____

____

____

Name ______________________________ Variable

Missing Numbers 2

Same shapes are same numbers.
Different shapes are different numbers.

□ + □ + ○ + ○ = 28

□ + □ = 18

1. What number is Square? __________
2. What number is Circle? __________
3. Tell how you found the numbers. ______________________

____

____

____

Name ______________________________

Missing Numbers 3

Same shapes are same numbers.
Different shapes are different numbers.

$$\square + \triangle + \square + \triangle = 20$$

$$\triangle + \triangle + \triangle + \triangle = 28$$

1. What number is Square? __________

2. What number is Triangle? __________

3. Tell how you found the numbers. ______________________

____

____

____

Name ______________________________

Variable

Missing Numbers 4

Same shapes are same numbers.
Different shapes are different numbers.

□ + □ + △ = 18

□ + △ = 10

1. What number is Square? __________

2. What number is Triangle? __________

3. Tell how you found the numbers. ____________________

____

____

____

Name ______________________________

■■■■□□
Variable

Missing Numbers 5

Same shapes are same numbers.
Different shapes are different numbers.

□ + □ + □ = 12

○ + □ + △ = 9

○ + □ = 6

1. What number is Square? ________

2. What number is Circle? ________

3. What number is Triangle? ________

4. Tell how you found the numbers. ________________

____

____

____

Name ______________________________

Missing Numbers 6

Same shapes are same numbers.
Different shapes are different numbers.

Square + Hexagon = 13

Triangle + Square + Triangle = 22

Square + Hexagon + Triangle = 20

1. What number is Square? ________
2. What number is Triangle? ________
3. What number is Hexagon? ________
4. Tell how you found the numbers. ____________________

____

____

____

Solutions

Missing Numbers 1

1. 6
2. 5
3. Possible answer: Both equations have a square and a circle. Replace the square and one circle in the first equation with 11. The new equation is 11 + Circle = 16. Circle is 5. In the second equation, if Circle is 5, Square is 11 − 5, or 6.

Missing Numbers 2

1. 9
2. 5
3. Possible answer: In the second equation, two squares are 18; one square is 9. Replace each square in the first equation with 9. The new equation is 9 + 9 + Circle + Circle = 28, so the circles total 28 − 18, or 10. Each circle is 5.

Missing Numbers 3

1. 3
2. 7
3. Possible answer: In the second equation, four triangles total 28. Each triangle is 28 ÷ 4, or 7. In the first equation, replace each triangle with 7. The squares total 20 − 7 − 7, or 6, so Square is 3.

Missing Numbers 4

1. 8
2. 2
3. Possible answer: In the second equation, Square and Triangle total 10. Replace one square and the triangle in the first equation with 10. That leaves one square equal to 18 − 10, or 8. In the second equation, replace Square with 8. Triangle is then 10 − 8, or 2.

Missing Numbers 5

1. 4
2. 2
3. 3
4. Possible answer: In the first equation, if three squares total 12, each square must be 12 ÷ 3, or 4. In the third equation, replace Square with 4. Circle equals 6 − 4, or 2. In the second equation, replace Square with 4 and Circle with 2. Triangle equals 9 − 4 − 2, or 3.

Missing Numbers 6

1. 8
2. 7
3. 5
4. Possible answer: In the first equation, Square and Hexagon total 13. In the third equation, replace Square and Hexagon with 13. Triangle is then equal to 20 − 13, or 7. In the second equation, replace each triangle with 7. That leaves Square equal to 22 − 7 − 7, or 8. In the first equation, if Square is 8, then Hexagon is 13 − 8, or 5.

Grid Sums

Goals

- Replace variables with numbers to make true equations.
- Make inferences.

Notes

In every problem in this set, there is at least one row or column that contains only one variable. Encourage students to identify these rows or columns as a first step. After students have completed the problems, have them replace each shape with its value and check that the values add to the row and column sums.

Solutions to all problems in this set appear on page 79.

Grid Sums 1

Questions to Ask

- What is the sum of the shapes in Row 1? (18)
- What is in Column 2? (a square, 5, and 3)
- What is the sum of Column 2? (14)
- To find the value of Square, which row or column would you look at first? (Row 1 or Column 2) Why? (You can figure Square's value directly. In Row 1, three squares total 18, so each square is 6. In Column 2, Square plus 8 totals 14, so Square is 14 − 8, or 6.)

Solutions

1. 6

2. 4

3. Possible answer: In Row 1, three squares total 18, so each square is 6. In Column 1, replace Square with 6. That means the two circles are equal to 14 − 6, or 8. Circle is 8 ÷ 2, or 4.

Name ____________________

Variable

Grid Sums 1

Same shapes are same numbers.
Different shapes are different numbers.
The number at the end of each row and each column is the sum.

Column

Row	1	2	3	
1	□	□	□	18
2	○	5	□	15
3	○	3	○	11
	14	14	16	

1. What number is Square? __________

2. What number is Circle? __________

3. Tell how you found the numbers. ____________________

____

____

____

Grid Sums 2

Same shapes are same numbers.
Different shapes are different numbers.
The number at the end of each row and each column is the sum.

Column

Row	1	2	3	
1	□	○	□	13
2	○	○	2	8
3	○	3	□	11
	11	9	12	

1. What number is Square? ________

2. What number is Circle? ________

3. Tell how you found the numbers. ________________________

____

____

____

Name ______________________

Variable

Grid Sums 3

Same shapes are same numbers.
Different shapes are different numbers.
The number at the end of each row and each column is the sum.

Column

Row	1	2	3	
1	□	□	○	26
2	2	○	□	19
3	□	9	□	27
	20	26	26	

1. What number is Square? __________

2. What number is Circle? __________

3. Tell how you found the numbers. ______________________

____

____

____

Name ______________________

Variable

Grid Sums 4

Same shapes are same numbers.
Different shapes are different numbers.
The number at the end of each row and each column is the sum.

Column

Row	1	2	3	
1	□	○	○	12
2	7	5	△	18
3	□	○	□	18
	23	9	16	

1. What number is Square? ________

2. What number is Triangle? ________

3. What number is Circle? ________

4. Tell how you found the numbers. ________________

____

____

____

Name ______________________________

■■■■□□ Variable

Grid Sums 5

Same shapes are same numbers.
Different shapes are different numbers.
The number at the end of each row and each column is the sum.

Column

Row	1	2	3	
1	□	8	7	17
2	△	5	○	10
3	△	○	□	7
	4	17	13	

1. What number is Square? __________
2. What number is Triangle? __________
3. What number is Circle? __________
4. Tell how you found the numbers. ______________________

____

____

____

Name ______________________________

■■■■□□
Variable

Grid Sums 6

Same shapes are same numbers.
Different shapes are different numbers.
The number at the end of each row and each column is the sum.

Column

Row	1	2	3	
1	□	○	△	14
2	2	△	□	10
3	5	3	□	11
	10	14	11	

1. What number is Square? __________
2. What number is Triangle? __________
3. What number is Circle? __________
4. Tell how you found the numbers. ______________________

____

____

____

Solutions

Grid Sums 1

1. 6
2. 4
3. Possible answer: In Row 1, three squares total 18, so each square is 6. In Column 1, replace Square with 6. That means the two circles are equal to $14 - 6$, or 8. Circle is $8 \div 2$, or 4.

Grid Sums 2

1. 5
2. 3
3. Possible answer: In Row 2, the two circles plus 2 total 8. The two circles are $8 - 2$, or 6, so each circle is 3. In Column 3, two squares plus 2 total 12, so two squares are $12 - 2$, or 10, and each square is 5.

Grid Sums 3

1. 9
2. 8
3. Possible answer: In Column 1, the two squares are equal to $20 - 2$, or 18, so Square is 9. In Row 1, if Square is 9, then Circle is equal to $26 - 9 - 9$, or 8.

Grid Sums 4

1. 8
2. 6
3. 2
4. Possible answer: In Column 1, two squares plus 7 total 23. Thus, two squares are $23 - 7$, or 16; Square is $16 \div 2$, or 8. In Column 2, two circles plus 5 total 9. So the circles are $9 - 5$, or 4; Circle is 2. In Row 2, Triangle is $18 - 5 - 7$, or 6.

Grid Sums 5

1. 2
2. 1
3. 4
4. Possible answer: In Row 1, Square is $17 - 8 - 7$, or 2. In Column 2, Circle is $17 - 8 - 5$, or 4. Replace Circle in Row 2 with 4. Then Triangle equals $10 - 5 - 4$, or 1.

Grid Sums 6

1. 3
2. 5
3. 6
4. Possible answer: In Column 1, Square equals $10 - 5 - 2$, or 3. Replace each square in Column 3 with 3. Triangle equals $11 - 3 - 3$, or 5. Replace Triangle in Column 2 with 5. That leaves Circle equal to $14 - 3 - 5$, or 6.

Bridge Sums

Goals

- Identify relationships among variables.
- Replace variables with numbers in systems of equations.
- Use substitution as a method for solving systems of equations.

Notes

To encourage students to experiment to figure out the number in each box, have them use chips or counters. Some students may question the fact that the sum of the numbers on the bridges is greater than the sum of the secret numbers. Help students to recognize that this occurs because each number is added twice. When the bridge numbers are added, (A + B) + (A + C) + (B + C), note that this simplifies to 2A + 2B + 2C. The sum of the bridge numbers is twice the sum of the secret numbers.

Solutions to all problems in this set appear on page 87.

Bridge Sums 1

Questions to Ask

- What is the sum of the secret numbers in boxes A, B, and C? (15)
- What numbers could A and B represent? (any pairs of whole numbers that add to 8: 0, 8; 1, 7; 2, 6; . . . ; 8, 0)
- If A is 6, what is the value of B? (2) If A is 6, what is the value of C? (8)
- Is it possible for B to be 1 and C to be 8? (No, B + C must be 10.)
- If you know that A + B is 8 and the sum of the secret numbers is 15, can you figure out the secret number in C? (yes) How? (If A + B + C is 15 and A + B is 8, then C is 15 − 8, or 7.)

Solutions

1. 5

2. 3

3. 7

4. Possible answer: Since A + B is 8, replace A + B with 8 in the equation A + B + C = 15. Then C = 15 − 8, or 7. In A + C = 12, since C is 7, A is 12 − 7, or 5. In A + B = 8, since A is 5, B is 8 − 5, or 3.

Name ____

■■■■□□ Variable

Bridge Sums 1

There is a secret number in each of the boxes A, B, and C. The number on each bridge is the sum of the secret numbers in the two boxes that are connected. The sum of all three secret numbers is 15.

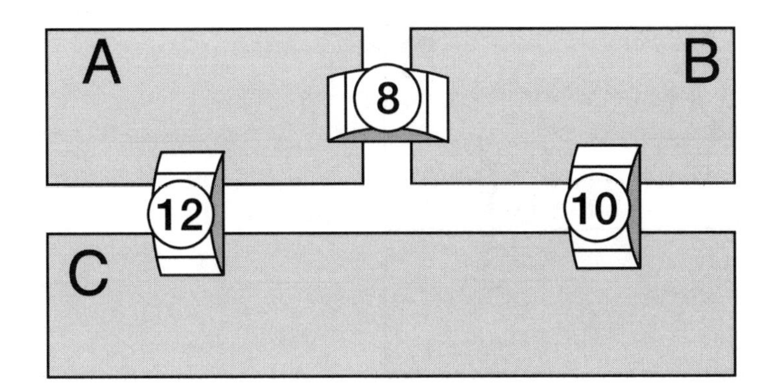

Use the equations to find the secret numbers.

$A + B = 8$ $\quad$ $B + C = 10$

$A + C = 12$ $\quad$ $A + B + C = 15$

1. A = __________

2. B = __________

3. C = __________

4. Tell how you figured out the secret number in each box.

____

____

____

Name ______________________________

■■■■□□
Variable

Bridge Sums 2

There is a secret number in each of the boxes A, B, and C. The number on each bridge is the sum of the secret numbers in the two boxes that are connected. The sum of all three secret numbers is 7.

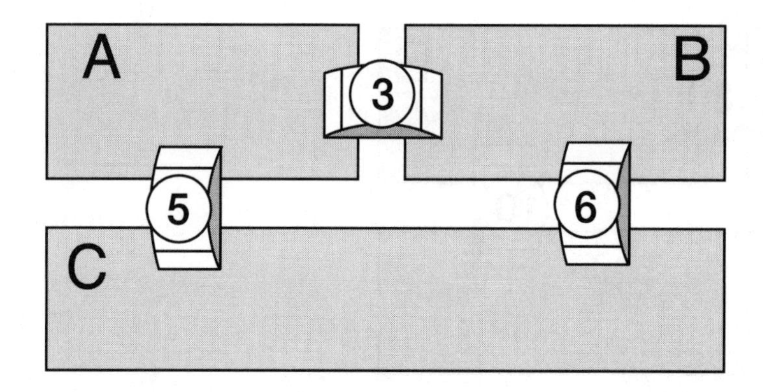

Use the equations to find the secret numbers.

$A + B = 3$ $\qquad$ $B + C = 6$

$A + C = 5$ $\qquad$ $A + B + C = 7$

1. A = __________

2. B = __________

3. C = __________

4. Tell how you figured out the secret number in each box.

____

____

____

Name ______________________________

Bridge Sums 3

There is a secret number in each of the boxes A, B, and C. The number on each bridge is the sum of the secret numbers in the two boxes that are connected. The sum of all three secret numbers is 18.

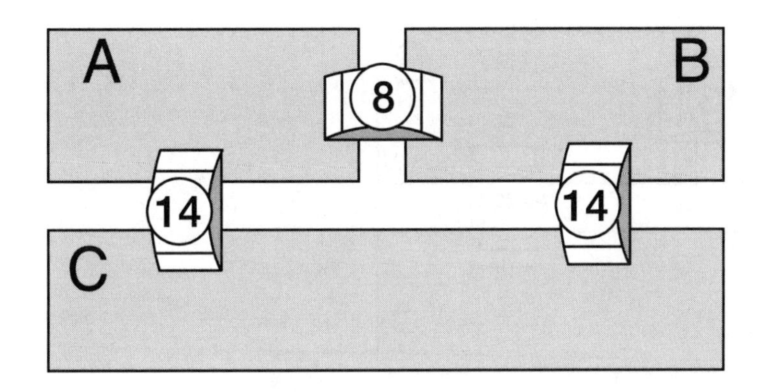

Use the equations to find the secret numbers.

$A + B = 8$ $B + C = 14$

$A + C = 14$ $A + B + C = 18$

1. A = __________

2. B = __________

3. C = __________

4. Tell how you figured out the secret number in each box.

____

____

____

Name ______________________________

Variable

Bridge Sums 4

There is a secret number in each of the boxes A, B, and C. The number on each bridge is the sum of the secret numbers in the two boxes that are connected. The sum of all three secret numbers is 21.

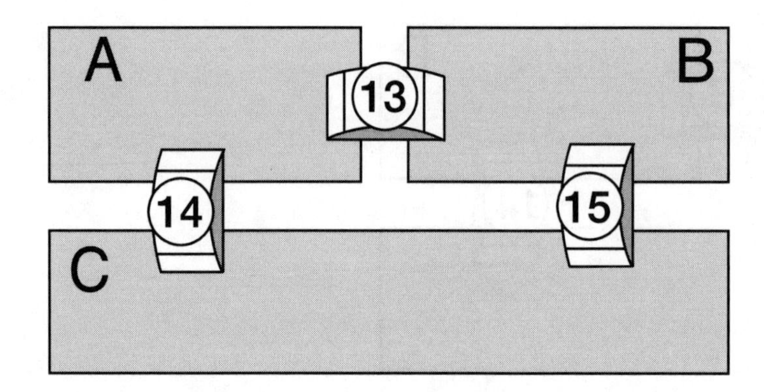

Use the equations to find the secret numbers.

$A + B = 13$ $\quad B + C = 15$

$A + C = 14$ $\quad A + B + C = 21$

1. A = __________

2. B = __________

3. C = __________

4. Tell how you figured out the secret number in each box.

____

____

____

Bridge Sums 5

There is a secret number in each of the boxes A, B, and C. The number on each bridge is the sum of the secret numbers in the two boxes that are connected. The sum of all three secret numbers is 24.

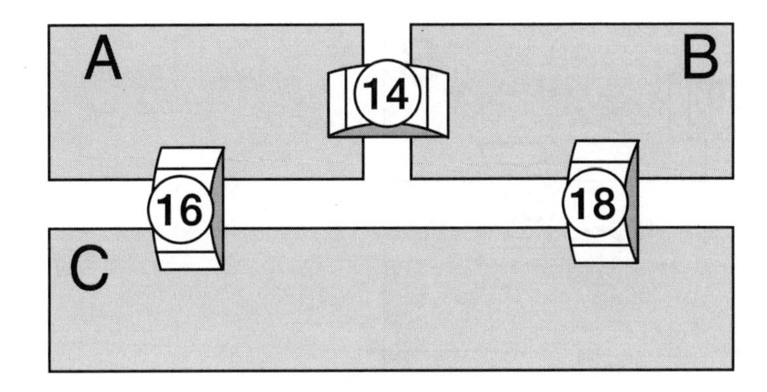

Use the equations to find the secret numbers.

$A + B = 14$ $B + C = 18$

$A + C = 16$ $A + B + C = 24$

1. A = ________

2. B = ________

3. C = ________

4. Tell how you figured out the secret number in each box.

____

____

____

Name ______________________________

■■■■□□
Variable

Bridge Sums 6

There is a secret number in each of the boxes A, B, and C. The number on each bridge is the sum of the secret numbers in the two boxes that are connected. The sum of all three secret numbers is 60.

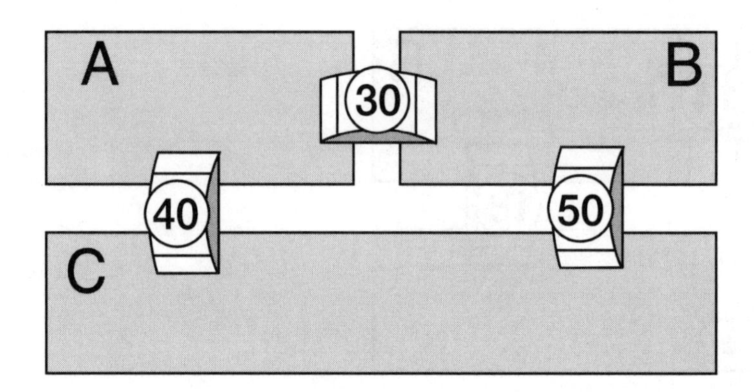

Use the equations to find the secret numbers.

$A + B = 30$ $B + C = 50$

$A + C = 40$ $A + B + C = 60$

1. A = ________
2. B = ________
3. C = ________
4. Tell how you figured out the secret number in each box.

____

____

____

Solutions

Bridge Sums 1

1. 5
2. 3
3. 7
4. Possible answer: Since A + B is 8, replace A + B with 8 in the equation A + B + C = 15. Then C = 15 − 8, or 7.
   In A + C = 12, since C is 7, A is 12 − 7, or 5.
   In A + B = 8, since A is 5, B is 8 − 5, or 3.

Bridge Sums 2

1. 1
2. 2
3. 4
4. Possible answer: Since A + B is 3, replace A + B with 3 in the equation A + B + C = 7. Then C = 7 − 3, or 4.
   In A + C = 5, since C is 4, A is 5 − 4, or 1.
   In A + B = 3, since A is 1, B is 3 − 1, or 2.

Bridge Sums 3

1. 4
2. 4
3. 10
4. Possible answer: Since A + C is 14, replace A + C with 14 in the equation A + B + C = 18. Then B = 18 − 14, or 4.
   In A + B = 8, since B is 4, A is 8 − 4, or 4.
   In A + C = 14, since A is 4, C is 14 − 4, or 10.

Bridge Sums 4

1. 6
2. 7
3. 8
4. Possible answer: Since A + C is 14, replace A + C with 14 in the equation A + B + C = 21. Then B = 21 − 14, or 7.
   In A + B = 13, since B is 7, A is 13 − 7, or 6.
   In B + C = 15, since B is 7, C is 15 − 7, or 8.

Bridge Sums 5

1. 6
2. 8
3. 10
4. Possible answer: Since A + B is 14, replace A + B with 14 in the equation A + B + C = 24. Then C = 24 − 14, or 10.
   In A + C = 16, since C is 10, A is 16 − 10, or 6.
   In A + B = 14, since A is 6, B is 14 − 6, or 8.

Bridge Sums 6

1. 10
2. 20
3. 30
4. Possible answer: Since B + C is 50, replace B + C with 50 in the equation A + B + C = 60. Then A = 60 − 50, or 10.
   In A + B = 30, since A is 10, then B = 30 − 10, or 20.
   In A + C = 40, since A is 10, C is 40 − 10, or 30.

Function Tables

Goals

- Describe a rule for finding the output if you know the input.
- Use inverse operations to compute inputs when given outputs.

Notes

Prior to doing this problem set, help students to make posters of number families that might be useful to know in recognizing patterns in the output numbers. These might include multiples of 2, 3, 4, 5, and 10; the square numbers 1, 4, 9, 16, 25, . . . ; the triangular numbers 1, 3, 6, 10, 15, . . . ; and the cubic numbers 1, 8, 27, ...

Solutions to all problems in this set appear on page 95.

Function Tables 1

Questions to Ask

- When the input is 1, what is the output? (11)
- When the output is 12, what is the input? (2)
- How do you find the output for any input number? (To find the output, you add 10 to the input.)

Solutions

1.

Input	Output
0	10
1	11
2	12
5	15
8	**18**
9	**19**
11	21
30	40
72	**82**
90	100

2. subtract 10 from the output to get the input

3. 76

Name ______________________________

Function

Function Tables 1

The table shows some inputs and some outputs.

Input	Output
0	10
1	11
2	12
5	15
8	
9	
	21
	40
72	
	100

1. Complete the table.
2. Use words to write a rule for finding the input if you know the output. ______________________________

____

3. If the output is 86, what is the input? __________

Name ______________________________

Function Tables 2

The table shows some inputs and some outputs.

Input	Output
4	0
5	1
6	2
8	
12	
	18
	21
	34
72	
	95

1. Complete the table.
2. Use words to write a rule for finding the input if you know the output. ______________________________
____
3. If the output is 106, what is the input? __________

Name ______________________________

Function Tables 3

The table shows some inputs and some outputs.

Input	Output
0	0
1	2
2	4
3	
5	
8	
	20
	24
	36
47	

1. Complete the table.
2. Use words to write a rule for finding the input if you know the output. ______________________________

____

3. If the output is 104, what is the input? __________

Function Tables 4

The table shows some inputs and some outputs.

Input	Output
0	0
1	3
2	6
3	9
4	12
	15
6	
	21
	30
	60

1. Complete the table.
2. Use words to write a rule for finding the input if you know the output. ______________________________

____
3. If the output is 120, what is the input? __________

Name ___

Function Tables 5

The table shows some inputs and some outputs.

Input	Output
0	0
2	1
4	2
6	3
8	4
	5
12	
	10
	15
	20

1. Complete the table.

2. Use words to write a rule for finding the input if you know the output. ___

___

___

___

3. If the output is 125, what is the input? ___

Function Tables 6

The table shows some inputs and some outputs.

Input	Output
0	1
1	11
2	21
3	31
4	41
	51
6	
	71
	91
	201

1. Complete the table.
2. Use words to write a rule for finding the input if you know the output. ______________________________

   ____
3. If the output is 121, what is the input? __________

Solutions

Function Tables 1

1.

Input	Output
0	10
1	11
2	12
5	15
8	18
9	19
11	21
30	40
72	82
90	100

2. subtract 10 from the output to get the input
3. 76

Function Tables 2

1.

Input	Output
4	0
5	1
6	2
8	4
12	8
22	18
25	21
38	34
72	68
99	95

2. add 4 to the output to get the input
3. 110

Function Tables 3

1.

Input	Output
0	0
1	2
2	4
3	6
5	10
8	16
10	20
12	24
18	36
47	94

2. divide the output by 2 to get the input
3. 52

Function Tables 4

1.

Input	Output
0	0
1	3
2	6
3	9
4	12
5	15
6	18
7	21
10	30
20	60

2. divide the output by 3 to get the input
3. 40

Function Tables 5

1.

Input	Output
0	0
2	1
4	2
6	3
8	4
10	5
12	6
20	10
30	15
40	20

2. multiply the output by 2 to get the input
3. 250

Function Tables 6

1.

Input	Output
0	1
1	11
2	21
3	31
4	41
5	51
6	61
7	71
9	91
20	201

2. Subtract 1 from the output, then divide the difference by 10 to get the input.
3. 12

Start to End

Goals

- Use inverse operations.
- Follow and complete sequences of computations.

Notes

Encourage students to do these problems in two steps and record the answer after each step. For example, for the first question in Problem 1, students will record $3 + 4 = 7$ and then $7 + 2 = 9$. Be sure that they do not misuse the equal sign to write $3 + 4 = 7 + 2 = 9$. For questions 4–6, have students record the inverse operations they use to find the input when given the output.

Solutions to all problems in this set appear on page 103.

Start to End 1

Questions to Ask

- Suppose the Start number is 0. What must you do to 0 first? (add 4 to get a sum of 4)
- What must you do to 4? (add 2 to 4 to get 6)
- What is the End number? (6)
- Suppose the End number is 8. Work backward. What number must you add to 2 to get 8? ($8 - 2$, or 6)
- Go backward again. If you get 6 after adding 4 to the Start number, what is the Start number? ($6 - 4$, or 2)

Solutions

1. 9
2. 13
3. 18
4. 13
5. 24
6. 93

Name ______________________________

Start to End 1

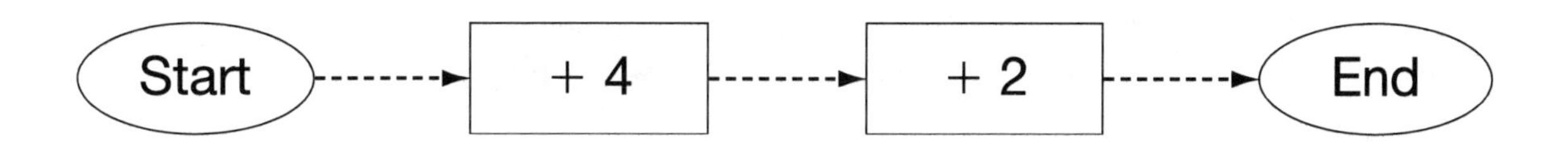

For each Start number, give the End number.

1. Start: 3 End: __________

2. Start: 7 End: __________

3. Start: 12 End: __________

For each End number, give the Start number.

4. End: 19 Start: __________

5. End: 30 Start: __________

6. End: 99 Start: __________

Name ______________________________

Start to End 2

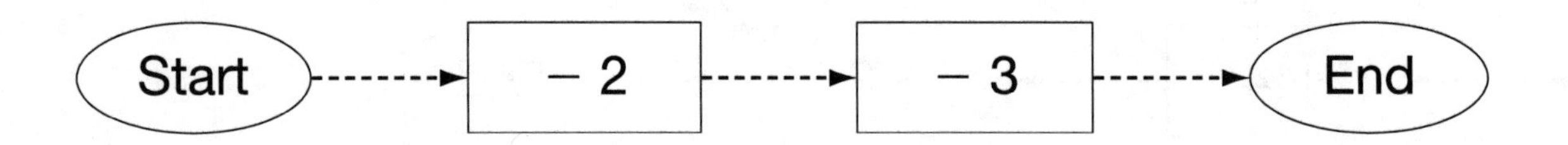

For each Start number, give the End number.

1. Start: 5 End: __________

2. Start: 8 End: __________

3. Start: 14 End: __________

For each End number, give the Start number.

4. End: 16 Start: __________

5. End: 27 Start: __________

6. End: 95 Start: __________

Name ______________________________

Start to End 3

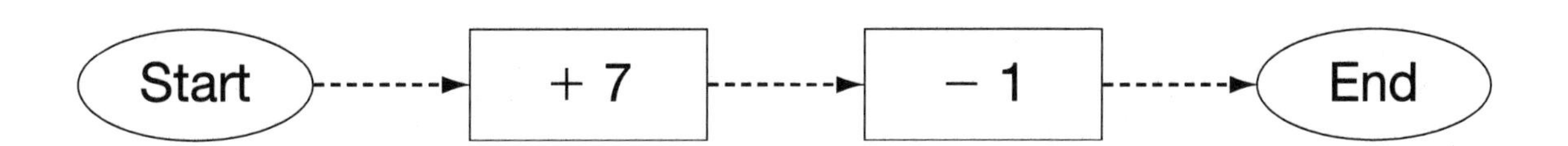

For each Start number, give the End number.

1. Start: 3 End: __________

2. Start: 6 End: __________

3. Start: 19 End: __________

For each End number, give the Start number.

4. End: 17 Start: __________

5. End: 29 Start: __________

6. End: 96 Start: __________

Name ______________________________

Start to End 4

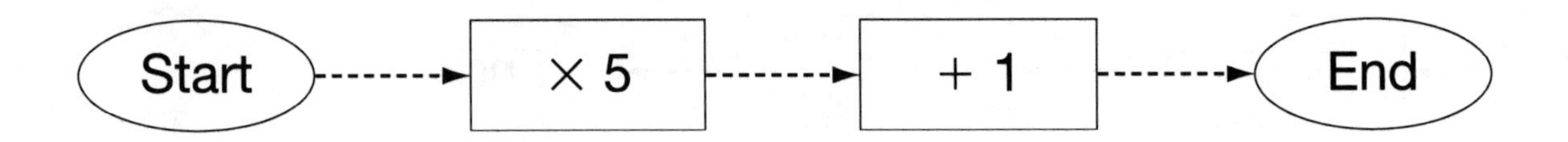

For each Start number, give the End number.

1. Start: 0 End: __________

2. Start: 4 End: __________

3. Start: 9 End: __________

For each End number, give the Start number.

4. End: 16 Start: __________

5. End: 26 Start: __________

6. End: 41 Start: __________

Name ____________________

Start to End 5

For each Start number, give the End number.

1. Start: 4 End: __________

2. Start: 24 End: __________

3. Start: 16 End: __________

For each End number, give the Start number.

4. End: 3 Start: __________

5. End: 9 Start: __________

6. End: 11 Start: __________

Name ______________________________

Start to End 6

Start → × 2 → + 2 → End

For each Start number, give the End number.

1. Start: 5 End: ________

2. Start: 7 End: ________

3. Start: 10 End: ________

For each End number, give the Start number.

4. End: 14 Start: ________

5. End: 20 Start: ________

6. End: 32 Start: ________

Solutions

Start to End 1

1. 9
2. 13
3. 18
4. 13
5. 24
6. 93

Start to End 2

1. 0
2. 3
3. 9
4. 21
5. 32
6. 100

Start to End 3

1. 9
2. 12
3. 25
4. 11
5. 23
6. 90

Start to End 4

1. 1
2. 21
3. 46
4. 3
5. 5
6. 8

Start to End 5

1. 0
2. 5
3. 3
4. 16
5. 40
6. 48

Start to End 6

1. 12
2. 16
3. 22
4. 6
5. 9
6. 15

Pattern Puzzles

Goals

- Identify and continue patterns.
- Write a rule relating number of chairs to number of tables.

Notes

If some students have difficulty identifying the patterns and formulating the rule, suggest that they copy the figures in the given pattern and then draw the next two or three figures. As they draw the figures, have them record their steps. This process may facilitate recognition of the nature of the change between consecutive figures and can lead to the formation of the rule.

Solutions to all problems in this set appear on page 111.

Pattern Puzzles 1

Questions to Ask

- What is the shape of the tabletops? (triangular)
- How many chairs are around one table? (three)
- When there are three tables, how many chairs are there? (five)
- How many chairs will there be for five tables? (seven)

Solutions

1. 7

2. 12

3. 22

4. Possible answers: To find the number of chairs, add 2 to the number of tables; $C = T + 2$.

Name ___________________________________

Pattern Puzzles 1

Each tabletop is a triangle.
There are chairs around the tables.
Imagine the pattern of tables and chairs continues.

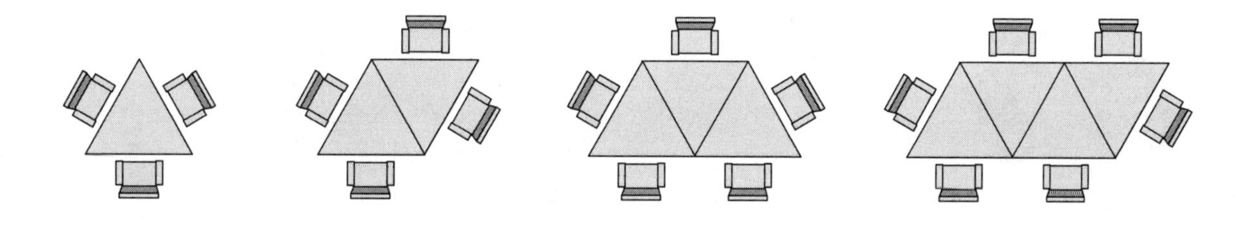

1. How many chairs are around 5 tables? __________
2. How many chairs are around 10 tables? __________
3. How many chairs are around 20 tables? __________
4. Write a rule for telling the number of chairs when you know the number of tables. ______________________________

____

Pattern Puzzles 2

Each tabletop is a triangle.
There are chairs around the tables.
Imagine this pattern of tables and chairs continues.

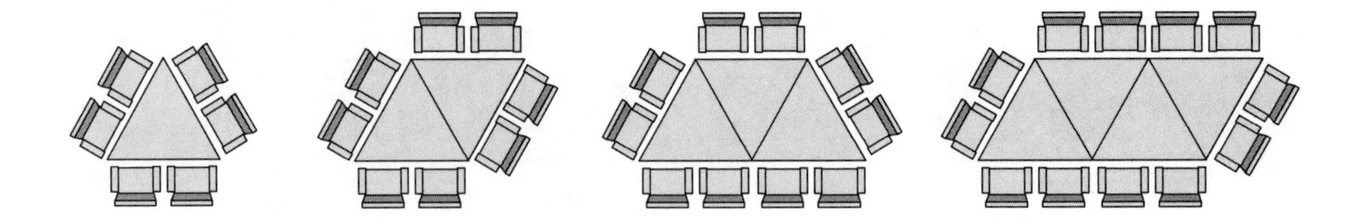

1. How many chairs are around 5 tables? __________
2. How many chairs are around 9 tables? __________
3. How many chairs are around 12 tables? __________
4. Write a rule for telling the number of chairs when you know the number of tables. ______________________________

____

Name ____

Pattern Puzzles 3

Each tabletop is a triangle.
There are chairs around the tables.
Imagine this pattern of tables and chairs continues.

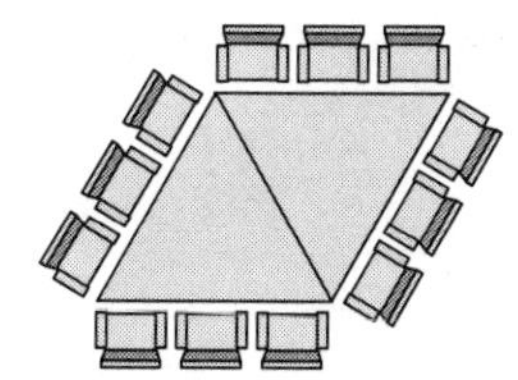
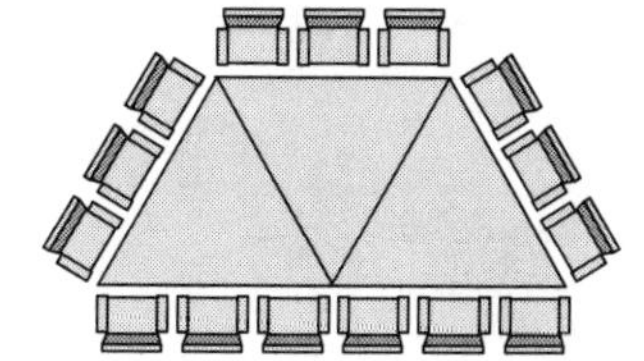
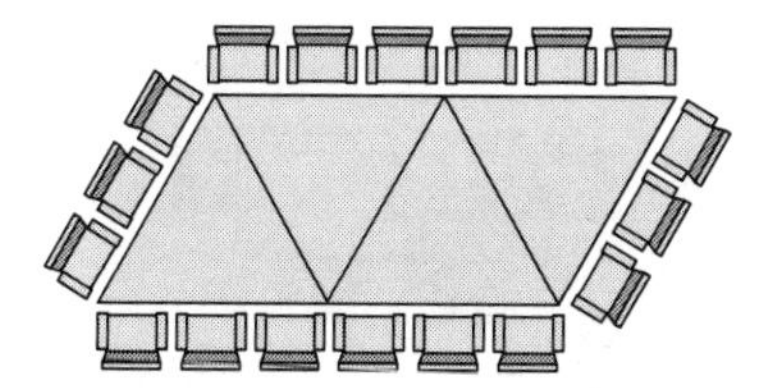

1. How many chairs are around 8 tables? __________
2. How many chairs are around 15 tables? __________
3. How many chairs are around 50 tables? __________
4. Write a rule for telling the number of chairs when you know the number of tables. ______________________________

____

Name ___

Pattern Puzzles 4

Each tabletop is a square.
There are chairs around the tables.
Imagine this pattern of tables and chairs continues.

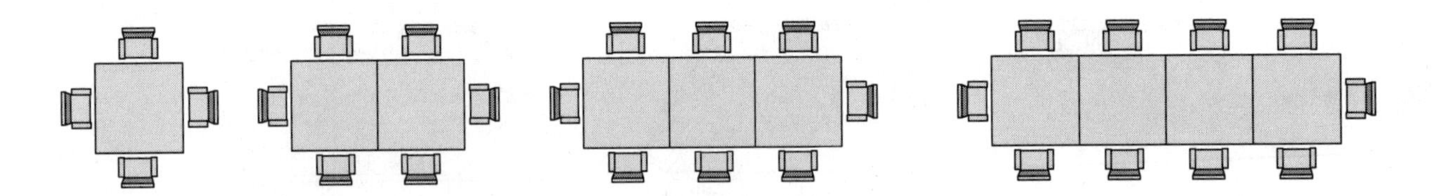

1. How many chairs are around 5 tables? ___
2. How many chairs are around 6 tables? ___
3. How many chairs are around 10 tables? ___
4. Write a rule for telling the number of chairs when you know the number of tables. ___

___

Name ______________________________

Pattern Puzzles 5

Each tabletop is a square.
There are chairs around the tables.
Imagine that this pattern of tables and chairs continues.

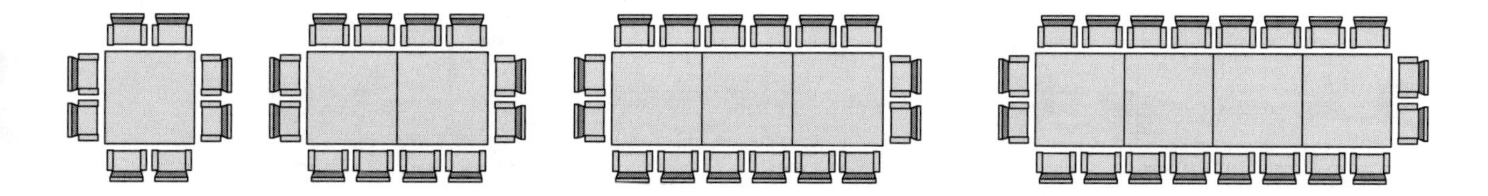

1. How many chairs are around 5 tables? ________
2. How many chairs are around 6 tables? ________
3. How many chairs are around 10 tables? ________
4. Write a rule for telling the number of chairs when you know the number of tables. ______________________________

____

Name ______________________

Pattern Puzzles 6

Each tabletop is a square.
There are chairs around the tables.
Imagine that this pattern of tables and chairs continues.

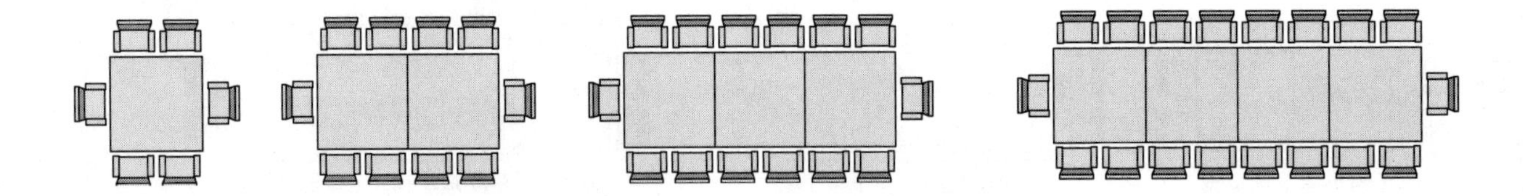

1. How many chairs are around 5 tables? __________
2. How many chairs are around 6 tables? __________
3. How many chairs are around 10 tables? __________
4. Write a rule for telling the number of chairs when you know the number of tables. ______________________________

____

Solutions

Pattern Puzzles 1

1. 7
2. 12
3. 22
4. Possible answers: To find the number of chairs, add 2 to the number of tables; $C = T + 2$.

Pattern Puzzles 2

1. 14
2. 22
3. 28
4. Possible answers: To find the number of chairs, add 2 to the number of tables. Then multiply the sum by 2; $C = (T + 2) \times 2$ or $C = 2T + 4$.

Pattern Puzzles 3

1. 30
2. 51
3. 156
4. Possible answers: To find the number of chairs, add 2 to the number of tables. Then multiply the sum by 3; $C = (T + 2) \times 3$ or $C = 3T + 6$.

Pattern Puzzles 4

1. 12
2. 14
3. 22
4. Possible answers: To find the number of chairs, multiply the number of tables by 2. Add 2 to the product; $C = 2T + 2$.

Pattern Puzzles 5

1. 24
2. 28
3. 44
4. Possible answers: To find the number of chairs, multiply the number of tables by 4 and add 4 to the product; $C = 4T + 4$.

Pattern Puzzles 6

1. 22
2. 26
3. 42
4. Possible answers: To find the number of chairs, multiply the number of tables by 4 and add 2 to the product; $C = 4T + 2$.

Lattice Logic

Goals

- Identify and generalize patterns.
- Identify inverse operations.
- Follow sequences of computations.

Notes

To enable students to follow the paths indicated by the arrows, suggest that they use different colors of pencils to trace the paths from number to number. To solve problems involving numbers that are not shown on the lattice, have students continue the lattice by recording numbers in the next three or more rows so that they can recognize the relationship between the direction of an arrow and the constant that is added or subtracted. Before students begin to follow the paths, encourage them to identify pairs of opposite direction arrows that "zero out." This will simplify the solution process.

Solutions to all problems in this set appear on page 119.

Lattice Logic 1

Questions to Ask

- What numbers are in Row 4? (31, 32, 33, 34, 35, . . ., 40)
- What numbers will be in Row 5? (41, 42, 43, 44, . . ., 50)
- What do the three dots above each column mean? (the pattern continues; there are more rows of numbers)
- In the example, $2\uparrow = 12$, what does the up-pointing arrow mean? (add 10)
- In the example, $18 \leftarrow = 17$, what does the left-pointing arrow mean? (subtract 1)

Solutions

1. 18
2. 14
3. 14
4. 56
5. 48
6. 6

Name ______________________________

Lattice Logic 1

This is a lattice of numbers. The dots mean the rows continue.

.	.	.	.	.	.	.	.	.	.	.
.	.	.	.	.	.	.	.	.	.	.
.	.	.	.	.	.	.	.	.	.	.
Row 4	31	32	33	34	35	36	37	38	38	40
Row 3	21	22	23	24	25	26	27	28	29	30
Row 2	11	12	13	14	15	16	17	18	19	20
Row 1	1	2	3	4	5	6	7	8	9	10
	A	**B**	**C**	**D**	**E**	**F**	**G**	**H**	**I**	**J**

Columns

Examples:

2↑ = 12

5 → = 6

35↓ = 25

18 ← = 17

12↑←← = 20

Complete the equations.

1. 8↑ = __________

2. 15 ← = __________

3. 24↓ = __________

4. 36↑ ↑ = __________

5. 47↑→↓ = __________

6. 6 ←↑→↓ = __________

Lattice Logic 2

This is a lattice of numbers. The dots mean the rows continue.

.	.	.	.	.	.	.	.	.	.	.
.	.	.	.	.	.	.	.	.	.	.
.	.	.	.	.	.	.	.	.	.	.
Row 4	31	32	33	34	35	36	37	38	38	40
Row 3	21	22	23	24	25	26	27	28	29	30
Row 2	11	12	13	14	15	16	17	18	19	20
Row 1	1	2	3	4	5	6	7	8	9	10
	A	**B**	**C**	**D**	**E**	**F**	**G**	**H**	**I**	**J**

Columns

Examples:
5↑↑ = 25
14↓↑→ = 15
19←←← = 16

Complete the equations.

1. 12↓← = __________
2. 18→↑↑ = __________
3. 39→↑← = __________
4. 7↑↑↑↑ = __________
5. 14←→↑↑ = __________
6. 52↓↓↓↑ = __________

Name ______________________________

Lattice Logic 3

This is a lattice of numbers. The dots mean the rows continue.

•	•	•	•	•	•	•	•	•	•	•
•	•	•	•	•	•	•	•	•	•	•
•	•	•	•	•	•	•	•	•	•	•
Row 4	31	32	33	34	35	36	37	38	38	40
Row 3	21	22	23	24	25	26	27	28	29	30
Row 2	11	12	13	14	15	16	17	18	19	20
Row 1	1	2	3	4	5	6	7	8	9	10
	A	B	C	D	E	F	G	H	I	J

Columns

Complete the equations.

1. 9↑→↑→↓ = __________
2. 64↓←↓←↓ = __________
3. 11↑→ → → → = __________
4. 23↓←↑ ↓ ↓ = __________
5. 105↑ ↑ ↑ ↓← → → = __________
6. 87↑ ↓→ ←↑→ = __________

Name ______________________________

Lattice Logic 4

This is a lattice of numbers. The dots mean the rows continue.

	A	B	C	D	E
.	.	.	.	.	.
.	.	.	.	.	.
.	.	.	.	.	.
Row 10	46	47	48	49	50
Row 9	41	42	43	44	45
Row 8	36	37	38	39	40
Row 7	31	32	33	34	35
Row 6	26	27	28	29	30
Row 5	21	22	23	24	25
Row 4	16	17	18	19	20
Row 3	11	12	13	14	15
Row 2	6	7	8	9	10
Row 1	1	2	3	4	5

Columns

Examples:

32↓↓= 22

32 →↑= 38

32↑← = 36

Complete the equations.

1. 17↑← = __________
2. 28↑←→↑← = __________
3. 43↑←↓→↑← = __________
4. 30↑←→↑←↓→↑← = __________
5. 48↑↑→→↓← = __________
6. 62 →→↓↓→ = __________

Lattice Logic 5

This is a lattice of numbers. The dots mean the rows continue.

.	.	.	.	.	.
.	.	.	.	.	.
.	.	.	.	.	.
Row 10	46	47	48	49	50
Row 9	41	42	43	44	45
Row 8	36	37	38	39	40
Row 7	31	32	33	34	35
Row 6	26	27	28	29	30
Row 5	21	22	23	24	25
Row 4	16	17	18	19	20
Row 3	11	12	13	14	15
Row 2	6	7	8	9	10
Row 1	1	2	3	4	5
	A	B	C	D	E

Columns

Complete the equations.

1. 73↑→→↓↓= __________
2. 85↓↓←←← = __________
3. 91→↓↓→ = __________
4. __________↑↑→ = 32
5. __________↑↑→→← = 39
6. __________↓↓→→→ = 30

Lattice Logic 6

This is a lattice of numbers. The dots mean the rows continue.

.	.	.	.	.	.
.	.	.	.	.	.
.	.	.	.	.	.
Row 10	46	47	48	49	50
Row 9	41	42	43	44	45
Row 8	36	37	38	39	40
Row 7	31	32	33	34	35
Row 6	26	27	28	29	30
Row 5	21	22	23	24	25
Row 4	16	17	18	19	20
Row 3	11	12	13	14	15
Row 2	6	7	8	9	10
Row 1	1	2	3	4	5
	A	B	C	D	E

Columns

Complete the equations.

1. 78↑↑→→↑ = __________
2. 98↓↓←←↑ = __________
3. __________↑↑→→ = 100
4. __________↓→→↓ = 89
5. __________→→↑↑→ = 95
6. __________↓↓↓←→→ = 79

Solutions

Lattice Logic 1

1. 18
2. 14
3. 14
4. 56
5. 48
6. 6

Lattice Logic 2

1. 1
2. 39
3. 49
4. 47
5. 34
6. 32

Lattice Logic 3

1. 21
2. 32
3. 25
4. 2
5. 126
6. 98

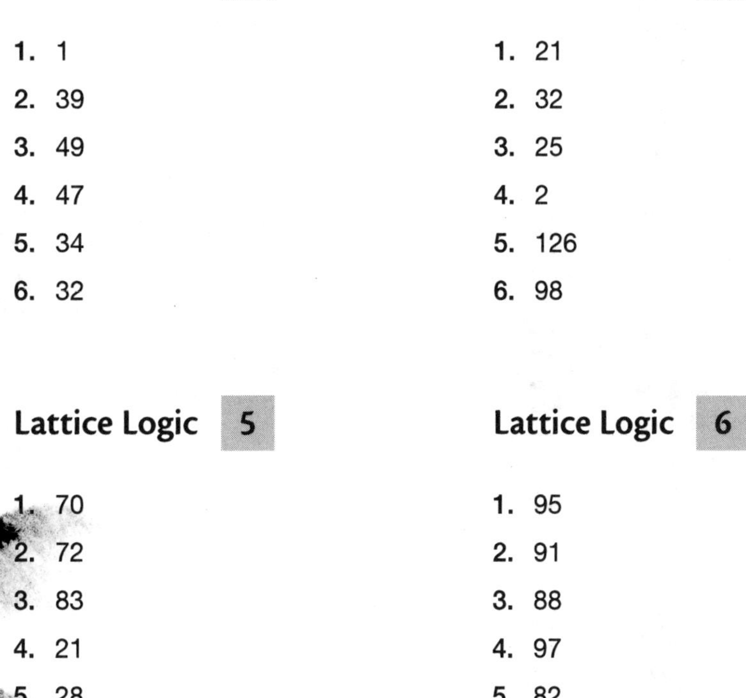

Lattice Logic 4

1. 21
2. 37
3. 47
4. 39
5. 54
6. 55

Lattice Logic 5

1. 70
2. 72
3. 83
4. 21
5. 28
6. 37

Lattice Logic 6

1. 95
2. 91
3. 88
4. 97
5. 82
6. 93

Certificate of Excellence

in Algebraic Thinking

This is to certify that

____

has satisfactorily completed all the problems for the big idea

____

and is [illegible] to be an expert.

Date __________ [illegible] ____________________

Grade __________ ____________________